Teacher Edition

Eureka Math
Grade 3
Module 2

Special thanks go to the Gordan A. Cain Center and to the Department of Mathematics at Louisiana State University for their support in the development of *Eureka Math*.

Published by Great Minds

Copyright © 2015 Great Minds. All rights reserved. No part of this work may be reproduced or used in any form or by any means — graphic, electronic, or mechanical, including photocopying or information storage and retrieval systems — without written permission from the copyright holder. "Great Minds" and "Eureka Math" are registered trademarks of Great Minds.

Printed in the U.S.A.
This book may be purchased from the publisher at eureka-math.org
10 9 8 7 6 5 4 3 2 1
ISBN 978-1-63255-364-5

A STORY OF UNITS

3 GRADE
Mathematics Curriculum

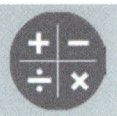

GRADE 3 • MODULE 2

Table of Contents
GRADE 3 • MODULE 2
Place Value and Problem Solving with Units of Measure

Module Overview 2

Topic A: Time Measurement and Problem Solving 10

Topic B: Measuring Weight and Liquid Volume in Metric Units 71

Mid-Module Assessment and Rubric 136

Topic C: Rounding to the Nearest Ten and Hundred 147

Topic D: Two- and Three-Digit Measurement Addition Using
the Standard Algorithm 182

Topic E: Two- and Three-Digit Measurement Subtraction Using
the Standard Algorithm 219

End-of-Module Assessment and Rubric 266

Answer Key 279

Grade 3 • Module 2
Place Value and Problem Solving with Units of Measure

OVERVIEW

In this 25-day module, students explore measurement using kilograms, grams, liters, milliliters, and intervals of time in minutes. Students begin by learning to tell and write time to the nearest minute using analog and digital clocks in Topic A (**3.MD.1**). They understand time as a continuous measurement through exploration with stopwatches, and use the number line, a continuous measurement model, as a tool for counting intervals of minutes within 1 hour (**3.MD.1**). Students see that an analog clock is a portion of the number line shaped into a circle. They use both the number line and clock to represent addition and subtraction problems involving intervals of minutes within 1 hour (**3.MD.1**).

Introduced in Topic B, kilograms and grams are measured using digital and spring scales. Students use manipulatives to build a kilogram and then decompose it to explore the relationship between the size and weight of kilograms and grams (**3.MD.2**). An exploratory lesson relates metric weight and liquid volume measured in liters and milliliters, highlighting the coherence of metric measurement. Students practice measuring liquid volume using the vertical number line and a graduated beaker (**3.MD.2**). Building on the estimation skills with metric length gained in Grade 2, students in Grade 3 use kilograms, grams, liters, and milliliters to estimate the weights and liquid volumes of familiar objects. Finally, they use their estimates to reason about solutions to one-step addition, subtraction, multiplication, and division word problems involving metric weight and liquid volume given in the same units (**3.MD.2**).

Now more experienced with measurement and estimation using different units and tools, students further develop their skills by learning to round in Topic C (**3.NBT.1**). They measure and then use place value understandings and the number line as tools to round two-, three-, and four-digit measurements to the nearest ten or hundred (**3.NBT.1**, **3.MD.1**, **3.MD.2**).

Students measure and round to solve problems in Topics D and E (**3.NBT.1**, **3.MD.1**, **3.MD.2**). In these topics, they use estimations to test the reasonableness of sums and differences precisely calculated using standard algorithms. From their work with metric measurement, students have a deeper understanding of the composition and decomposition of units. They demonstrate this understanding in every step of the addition and subtraction algorithms with two- and three-digit numbers, as 10 units are changed for 1 larger unit or 1 larger unit is changed for 10 smaller units (**3.NBT.2**). Both topics end in problem solving involving metric units or intervals of time. Students round to estimate and then calculate precisely using the standard algorithm to add or subtract two- and three-digit measurements given in the same units (**3.NBT.1**, **3.NBT.2**, **3.MD.1**, **3.MD.2**).

Notes on Pacing for Differentiation

If pacing is a challenge, consider the following modifications and omissions.

Omit Lesson 1. Prior to Lesson 2, use a stopwatch to time different activities such as lining up or moving to the meeting area of the classroom. Discuss the continuity of time. Reduce Lesson 2 by replacing the Minute Counting fluency with an activity in which students draw and label 14-centimeter number lines with tick marks at every centimeter in preparation for the Concept Development component of the lesson.

Omit Lesson 4, the first of two consecutive problem solving lessons involving time. Take note of the word problem analysis it provides, and consider embedding that work into the delivery of Lesson 5.

Consolidate Lessons 15 and 16. Within the lesson that results, include some problems that require regrouping once to add and some problems that require regrouping twice.

Consolidate Lessons 18 and 19. Within the lesson that results, include some problems that require regrouping once to subtract and some problems that require regrouping twice.

Omit Lesson 20. While it engages students in a study of estimation and provides practice with reasoning about the relationships between quantities, the lesson does not present new skills.

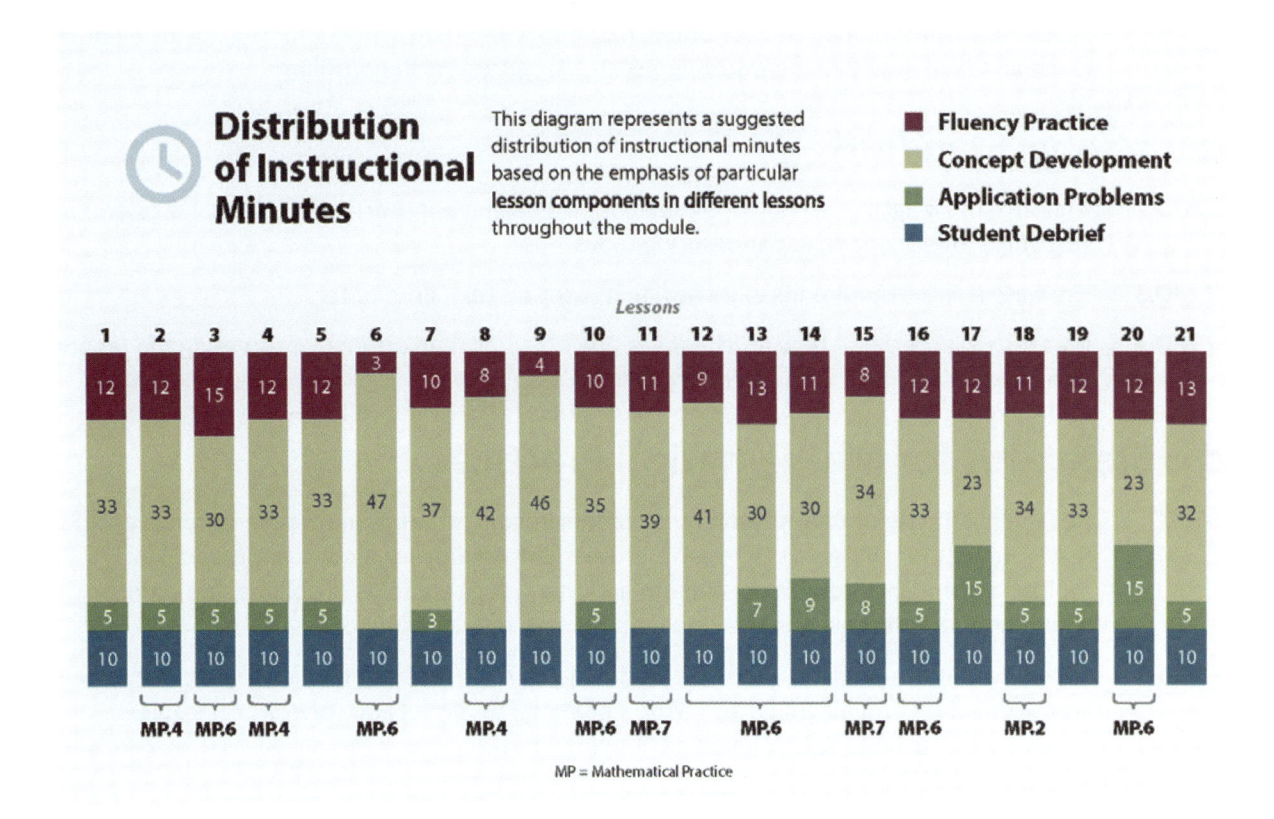

Distribution of Instructional Minutes

This diagram represents a suggested distribution of instructional minutes based on the emphasis of particular lesson components in different lessons throughout the module.

- ■ Fluency Practice
- ■ Concept Development
- ■ Application Problems
- ■ Student Debrief

Lesson	Fluency	Concept	Application	Debrief
1	12	33	5	10
2	12	33	5	10
3	15	30	5	10
4	12	33	5	10
5	12	33	5	10
6	3	47	—	10
7	10	37	3	10
8	8	42	—	10
9	4	46	—	10
10	10	35	5	10
11	11	39	—	10
12	9	41	—	10
13	13	30	7	10
14	11	30	9	10
15	8	34	8	10
16	12	33	5	10
17	12	23	15	10
18	11	34	5	10
19	12	33	5	10
20	12	23	15	10
21	13	32	5	10

Groupings: MP.4 (3), MP.6 (4), MP.4 (5), MP.6 (6), MP.4 (7–8), MP.6 (10), MP.7 (11), MP.6 (12–14), MP.7 (15), MP.6 (16), MP.2 (17–18), MP.6 (19–21)

MP = Mathematical Practice

Focus Grade Level Standards

Use place value understanding and properties of operations to perform multi-digit arithmetic.[1]

3.NBT.1 Use place value understanding to round whole numbers to the nearest 10 or 100.

3.NBT.2 Fluently add and subtract within 1000 using strategies and algorithms based on place value, properties of operations, and/or the relationship between addition and subtraction.

Solve problems involving measurement and estimation of intervals of time, liquid volumes, and masses of objects.

3.MD.1 Tell and write time to the nearest minute and measure time intervals in minutes. Solve word problems involving addition and subtraction of time intervals in minutes, e.g., by representing the problem on a number line diagram.

3.MD.2 Measure and estimate liquid volumes and masses of objects using standard units of grams (g), kilograms (kg), and liters (l). Add, subtract, multiply, or divide to solve one-step word problems involving masses or volumes that are given in the same units, e.g., by using drawings (such as a beaker with a measurement scale) to represent the problem.

Foundational Standards

2.MD.1 Measure the length of an object by selecting and using appropriate tools such as rulers, yardsticks, meter sticks, and measuring tapes.

2.MD.3 Estimate lengths using units of inches, feet, centimeters, and meters.

2.MD.4 Measure to determine how much longer one object is than another, expressing the length difference in terms of a standard length unit.

Focus Standards for Mathematical Practice

MP.2 **Reason abstractly or quantitatively.** Students decontextualize metric measurements and time intervals in minutes as they solve problems involving addition, subtraction, and multiplication. They round to estimate, and then precisely solve problems, evaluating solutions with reference to units and with respect to real-world contexts.

MP.4 **Model with mathematics.** Students model measurements on the place value chart. They create drawings and diagrams and write equations to model and solve word problems involving metric units and intervals of time in minutes.

---

[1] 3.NBT.3 is taught in Module 3.

| | A STORY OF UNITS | | Module Overview | 3•2 |

MP.6 **Attend to precision.** Students round to estimate sums and differences and then use the standard algorithms for addition and subtraction to calculate. They reason about the precision of their solutions by comparing estimations with calculations and by attending to specific units of measure.

MP.7 **Look for and make use of structure.** Students model measurements on the place value chart. Through modeling, they relate different units of measure and analyze the multiplicative relationship of the base ten system.

Overview of Module Topics and Lesson Objectives

Standards		Topics and Objectives		Days
3.NBT.2 3.MD.1	A	**Time Measurement and Problem Solving**		5
		Lesson 1:	Explore time as a continuous measurement using a stopwatch.	
		Lesson 2:	Relate skip-counting by fives on the clock and telling time to a continuous measurement model, the number line.	
		Lesson 3:	Count by fives and ones on the number line as a strategy to tell time to the nearest minute on the clock.	
		Lesson 4:	Solve word problems involving time intervals within 1 hour by counting backward and forward using the number line and clock.	
		Lesson 5:	Solve word problems involving time intervals within 1 hour by adding and subtracting on the number line.	
3.NBT.2 3.MD.2	B	**Measuring Weight and Liquid Volume in Metric Units**		6
		Lesson 6:	Build and decompose a kilogram to reason about the size and weight of 1 kilogram, 100 grams, 10 grams, and 1 gram.	
		Lesson 7:	Develop estimation strategies by reasoning about the weight in kilograms of a series of familiar objects to establish mental benchmark measures.	
		Lesson 8:	Solve one-step word problems involving metric weights within 100 and estimate to reason about solutions.	
		Lesson 9:	Decompose a liter to reason about the size of 1 liter, 100 milliliters, 10 milliliters, and 1 milliliter.	
		Lesson 10:	Estimate and measure liquid volume in liters and milliliters using the vertical number line.	
		Lesson 11:	Solve mixed word problems involving all four operations with grams, kilograms, liters, and milliliters given in the same units.	
		Mid-Module Assessment: Topics A–B (assessment ½ day, return ½ day, remediation or further applications 1 day)		2

Standards	Topics and Objectives		Days
3.NBT.1 3.MD.1 3.MD.2	C	**Rounding to the Nearest Ten and Hundred** Lesson 12: Round two-digit measurements to the nearest ten on the vertical number line. Lesson 13: Round two- and three-digit numbers to the nearest ten on the vertical number line. Lesson 14: Round to the nearest hundred on the vertical number line.	3
3.NBT.2 3.NBT.1 3.MD.1 3.MD.2	D	**Two- and Three-Digit Measurement Addition Using the Standard Algorithm** Lesson 15: Add measurements using the standard algorithm to compose larger units once. Lesson 16: Add measurements using the standard algorithm to compose larger units twice. Lesson 17: Estimate sums by rounding and apply to solve measurement word problems.	3
3.NBT.2 3.NBT.1 3.MD.1 3.MD.2	E	**Two- and Three-Digit Measurement Subtraction Using the Standard Algorithm** Lesson 18: Decompose once to subtract measurements including three-digit minuends with zeros in the tens or ones place. Lesson 19: Decompose twice to subtract measurements including three-digit minuends with zeros in the tens and ones places. Lesson 20: Estimate differences by rounding and apply to solve measurement word problems. Lesson 21: Estimate sums and differences of measurements by rounding, and then solve mixed word problems.	4
		End-of-Module Assessment: Topics A–E (assessment ½ day, return ½ day, remediation or further applications 1 day)	2
Total Number of Instructional Days			**25**

Terminology

New or Recently Introduced Terms and Symbols

- About (with reference to rounding and estimation, an answer that is not precise)
- Addend (the numbers that are added together in an addition equation, e.g., in 4 + 5, the numbers 4 and 5 are the addends)
- Capacity (the amount of liquid that a particular container can hold)
- Continuous (with reference to time as a continuous measurement)
- Endpoint[2] (used with rounding on the number line; the numbers that mark the beginning and end of a given interval)
- Gram (g, unit of measure for weight)
- Interval (time passed or a segment on the number line)
- Halfway (with reference to a number line, the midpoint between two numbers, e.g., 5 is halfway between 0 and 10)
- Kilogram (kg, unit of measure for mass)
- Liquid volume (the space a liquid takes up)
- Liter (L, unit of measure for liquid volume)
- Milliliter (mL, unit of measure for liquid volume)
- Plot (locate and label a point on a number line)
- Point (a specific location on the number line)
- Reasonable (with reference to how plausible an answer is, e.g., "Is your answer reasonable?")
- Round[3] (estimate a number to the nearest 10 or 100 using place value)
- Second (a unit of time)
- Standard algorithm (for addition and subtraction)
- ≈ (symbol used to show that an answer is approximate)

Familiar Terms and Symbols[4]

- Analog clock (a clock that is not digital)
- Centimeter (cm, unit of measurement)
- Compose (change 10 smaller units for 1 of the next larger unit on the place value chart)
- Divide (e.g., 4 ÷ 2 = 2)
- Estimate (approximation of the value of a quantity or number)

---

[2] Originally introduced in Grade 2, but treated as new vocabulary in this module.
[3] Originally introduced in Grade 2, but treated as new vocabulary in this module.
[4] These are terms and symbols students have used or seen previously.

- Horizontal (with reference to how an equation is written, e.g., 3 + 4 = 7 is written horizontally)
- Measure (a quantity representing a weight or liquid volume or the act of finding the size or amount of something)
- Mental math (calculations performed in one's head, without paper and pencil)
- Meter (m, unit of measurement)
- Minute (a unit of time)
- Multiply (e.g., 2 × 2 = 4)
- Number line (may be vertical or horizontal; vertical number line shown on the next page)
- Rename (regroup units, e.g., when solving with the standard algorithm)
- Simplifying strategy (transitional strategies that move students toward mental math, e.g., *make ten* to add 7 and 6, ((7 + 3) + 3 = 13))
- Unbundle (regroup units, e.g., in the standard algorithm)
- Vertical (with reference to how an equation is written; equations solved using the standard algorithm are typically written vertically)

Suggested Tools and Representations

- Beaker (100 mL and optional 1 liter)
- Beans (e.g., pinto beans, used for making benchmark baggies at different weights)
- Bottles (empty, plastic, labels removed, measuring 2 liters; 1 for every group of 3 students)
- Clocks (analog and digital)
- Containers (clear plastic, 1 each: cup, pint, quart, gallon)
- Cups (16, clear plastic, with capacity of about 9 oz)
- Cylinder (a slim, cylindrical container whose sides are marked with divisions or units of measure)
- Dropper (for measuring 1 mL)
- Liter-sized container (a container large enough to hold and measure 1 liter)
- Meter strip (e.g., meter stick)
- Pan balance (pictured to the right)
- Pitchers (plastic, 1 for each group of 3 students)
- Place value cards (pictured to the right)
- Place value chart and disks (pictured to the right)
- Place value disks (pictured to the right)
- Popcorn kernels (enough to make baggies weighing 36 g per student pair)
- Rice (e.g., white rice, used for making benchmark baggies at different weights)

Place Value Cards

Spring Scale

Pan Balance

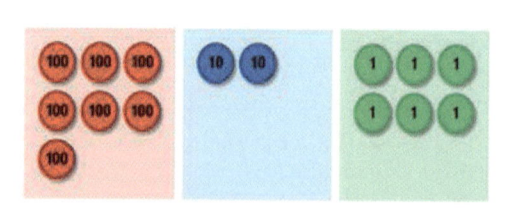

Sample place value chart without headings. Place value disks are shown in each column.

- Ruler (measuring centimeters)
- Scales (digital and spring, measures the mass of an object in grams)
- Sealable plastic bags (gallon-sized and sandwich-sized for making benchmark baggies)
- Stopwatch (handheld timepiece that measures time elapsed from when activated to when deactivated, 1 per student pair)
- Tape diagram (method for modeling)
- Ten-frame (pictured to the right)
- Vertical number line (pictured to the right)
- Weights (1 set per student pair: 1 g, 10 g, 100 g, 1 kg, or premeasured and labeled bags of rice or beans)

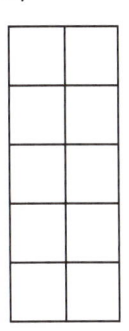

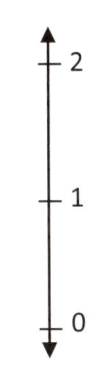

Ten-Frame   Vertical Number Line

Scaffolds[5]

The scaffolds integrated into *A Story of Units* give alternatives for how students access information as well as express and demonstrate their learning. Strategically placed margin notes are provided within each lesson elaborating on the use of specific scaffolds at applicable times. They address many needs presented by English language learners, students with disabilities, students performing above grade level, and students performing below grade level. Many of the suggestions are organized by Universal Design for Learning (UDL) principles and are applicable to more than one population. To read more about the approach to differentiated instruction in *A Story of Units,* please refer to "How to Implement *A Story of Units*."

Assessment Summary

Type	Administered	Format	Standards Addressed
Mid-Module Assessment Task	After Topic B	Constructed response with rubric	3.NBT.2 3.MD.1 3.MD.2
End-of-Module Assessment Task	After Topic E	Constructed response with rubric	3.NBT.1 3.NBT.2 3.MD.1 3.MD.2 3.OA.7[6]

---

[5] Students with disabilities may require Braille, large print, audio, or special digital files. Please visit the website www.p12.nysed.gov/specialed/aim for specific information on how to obtain student materials that satisfy the National Instructional Materials Accessibility Standard (NIMAS) format.

[6] Although 3.OA.7 is not a focus standard in this module, it does represent the major fluency for Grade 3. Module 2 fluency instruction provides systematic practice for maintenance and growth. The fluency page on the End-of-Module Assessment directly builds on the assessment given at the end of Module 1 and leads into the assessment that will be given at the end of Module 3.

A STORY OF UNITS

GRADE 3

Mathematics Curriculum

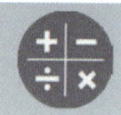

GRADE 3 • MODULE 2

Topic A

Time Measurement and Problem Solving

3.NBT.2, 3.MD.1

Focus Standards:	3.NBT.2	Fluently add and subtract within 1000 using strategies and algorithms based on place value, properties of operations, and/or the relationship between addition and subtraction.
	3.MD.1	Tell and write time to the nearest minute and measure time intervals in minutes. Solve word problems involving addition and subtraction of time intervals in minutes, e.g., by representing the problem on a number line diagram.
Instructional Days:	5	
Coherence -Links from:	G2–M2	Addition and Subtraction of Length Units
-Links to:	G4–M2	Unit Conversions and Problem Solving with Metric Measurement

Lesson 1 is an exploration in which students use stopwatches to measure time as a physical quantity. They might, for example, time how long it takes to write the fact 7 × 8 = 56 forty times or measure how long it takes to write numbers from 0 to 100. Students time their own segments as they run a relay, exploring the continuity of time by contextualizing their small segment within the number of minutes it takes the whole team to run.

Lesson 2 builds students' understanding of time as a continuous unit of measurement. This lesson draws upon the Grade 2 skill of telling time to the nearest 5 minutes (**2.MD.7**) and the multiplication learned in Module 1 as students relate skip-counting by fives and telling time to the number line. They learn to draw the model, labeling hours as endpoints and multiples of 5 (shown below). Through this work, students recognize the analog clock as a portion of the number line shaped into a circle and, from this point on, use the number line as a tool for modeling and solving problems.

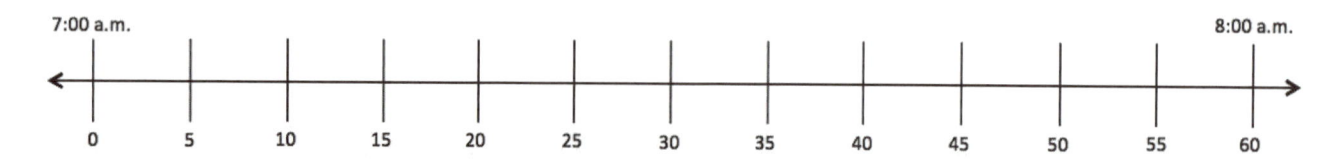

7:00 a.m.   8:00 a.m.

```
←|——|——|——|——|——|——|——|——|——|——|——|——|→
  0   5   10  15  20  25  30  35  40  45  50  55  60
```

Topic A

Lesson 3 increases students' level of precision as they read and write time to the nearest minute. Students draw number line models that represent the minutes between multiples of 5 (number line model shown below). They quickly learn to apply the strategy of counting by fives and some ones to read time to the nearest minute on the clock. In preparation for Lessons 4 and 5, students add minutes by counting on the number line and clock. For example, they might use the *count by fives and some ones* strategy to locate 17 minutes and then keep counting to find 4 minutes more.

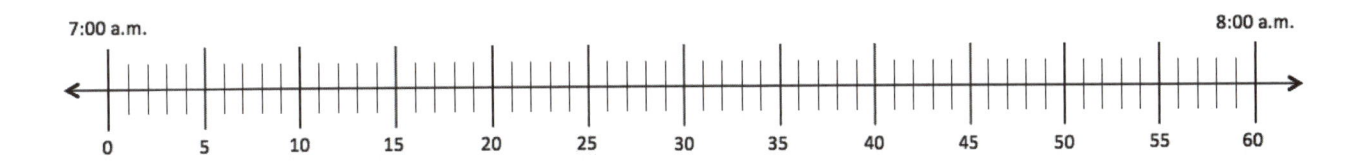

In Lesson 4, students begin measuring time intervals in minutes within 1 hour to solve word problems. They reinforce their understanding of time as a continuous unit of measurement by counting forward and backward using the number line and the clock. They might solve, for example, a problem such as, "Beth leaves her house at 8:05 a.m. and arrives at school at 8:27 a.m. How many minutes does Beth spend traveling to school?"

Lesson 5 carries problem solving with time a step further. Students measure minute intervals and then add and subtract the intervals to solve problems. Students might solve problems such as, "I practiced the piano for 25 minutes and the clarinet for 30 minutes. How long did I spend practicing my instruments?" Calculations with time in this lesson—and throughout Grade 3—never cross over an hour or involve students converting between hours and minutes.

A Teaching Sequence Toward Mastery of Time Measurement and Problem Solving

Objective 1: Explore time as a continuous measurement using a stopwatch.
(Lesson 1)

Objective 2: Relate skip-counting by fives on the clock and telling time to a continuous measurement model, the number line.
(Lesson 2)

Objective 3: Count by fives and ones on the number line as a strategy to tell time to the nearest minute on the clock.
(Lesson 3)

Objective 4: Solve word problems involving time intervals within 1 hour by counting backward and forward using the number line and clock.
(Lesson 4)

Objective 5: Solve word problems involving time intervals within 1 hour by adding and subtracting on the number line.
(Lesson 5)

Lesson 1

Objective: Explore time as a continuous measurement using a stopwatch.

Suggested Lesson Structure

- ■ Fluency Practice   (12 minutes)
- ■ Application Problem   (5 minutes)
- ■ Concept Development   (33 minutes)
- ■ Student Debrief   (10 minutes)
    Total Time   **(60 minutes)**

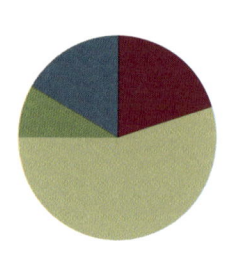

Fluency Practice (12 minutes)

- Tell Time on the Clock **2.MD.7**   (3 minutes)
- Minute Counting **3.MD.1**   (6 minutes)
- Group Counting **3.OA.1**   (3 minutes)

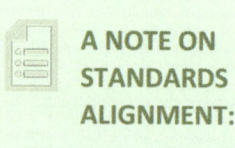

> **A NOTE ON STANDARDS ALIGNMENT:**
>
> In this lesson, students use stopwatches to measure time. To understand how to use a stopwatch and to begin to conceptualize time as a continuous measurement, students need some familiarity with seconds. The introduction of seconds anticipates Grade 4 content (**4.MD.1**).
>
> Seconds are used as a unit in the Application Problem and also as a unit of measure that students explore in Part 1 of the lesson as they familiarize themselves with stopwatches.

Tell Time on the Clock (3 minutes)

Materials: (T) Analog clock for demonstration (S) Personal white board

Note: This activity reviews the Grade 2 standard of telling and writing time to the nearest 5 minutes. It prepares students to count by 5-minute intervals on the number line and clock in Lesson 2.

- T: (Show an analog demonstration clock.) Start at 12 and count by 5 minutes on the clock. (Move finger from 12 to 1, 2, 3, 4, etc., as students count.)
- S: 5, 10, 15, 20, 25, 30, 35, 40, 45, 50, 55, 60.
- T: I'll show a time on the clock. Write the time on your personal white board. (Show 11:10.)
- S: (Write 11:10.)
- T: (Show 6:30.)
- S: (Write 6:30.)

Repeat the process, varying the hour and 5-minute interval so that students read and write a variety of times to the nearest 5 minutes.

Minute Counting (6 minutes)

Note: This activity reviews the Grade 2 standard of telling and writing time to the nearest 5 minutes. It prepares students to count by 5-minute intervals on the number line and clock in Lesson 2. Students also practice group counting strategies for multiplication in the context of time.

- T: There are 60 minutes in 1 hour. Count by 5 minutes to 1 hour.
- S: 5 minutes, 10 minutes, 15 minutes, 20 minutes, 25 minutes, 30 minutes, 35 minutes, 40 minutes, 45 minutes, 50 minutes, 55 minutes, 60 minutes. (Underneath 60 minutes, write 1 hour.)
- T: How many minutes are in a half hour?
- S: 30 minutes.
- T: Count by 5 minutes to 1 hour. This time, say *half hour* when you get to 30 minutes.

Repeat the process using the following suggested sequence:

- Count by 10 minutes and 6 minutes to 1 hour.
- Count by 3 minutes to a half hour.

Group Counting (3 minutes)

Note: Group counting reviews interpreting multiplication as repeated addition. Counting by sevens, eights, and nines in this activity anticipates multiplication using those units in Module 3.

Direct students to count forward and backward using the following suggested sequence, occasionally changing the direction of the count:

- Sevens to 28
- Eights to 32
- Nines to 36

Application Problem (5 minutes)

Ms. Bower helps her kindergartners tie their shoes. It takes her 5 seconds to tie 1 shoe. How many seconds does it take Ms. Bower to tie 8 shoes?

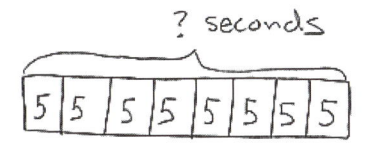

$8 \times 5 \text{ seconds} = 40 \text{ seconds}$

It takes Ms. Bower 40 seconds to tie 8 shoes.

A NOTE ON STANDARDS ALIGNMENT:

Seconds exceed the standard for Grade 3, which expects students to tell time to the nearest minute. The standards introduce seconds in Grade 4 (**4.MD.1**).

Note: This reviews multiplication from Module 1 and gets students thinking about how long it takes to complete an activity or task. It leads into the Concept Development by previewing the idea of seconds as a unit of time.

A STORY OF UNITS   Lesson 1   3•2

Concept Development (33 minutes)

Materials: (T) Stopwatch and classroom clock (S) Stopwatch, personal white board

Part 1: Explore seconds as a unit of time.

T: It takes Ms. Bower 5 seconds to tie one shoe. Does it take a very long time to tie a shoe?

S: No!

T: Let's see how long a second is. (Let the stopwatch tick off a second.)

T: It's a short amount of time! Let's see how long 5 seconds is so we know how long it takes Ms. Bower to tie 1 shoe. (Let the stopwatch go for 5 seconds.)

T: Let's see how long 40 seconds lasts. That's the amount of time it takes Ms. Bower to tie 8 shoes. (Let the stopwatch go for 40 seconds.) Tell the count after every 5 seconds.

S: (Watch the stopwatch.) 5. 10. 15. 20. 25. 30. 35. 40.

T: **Seconds** are a unit of time. They're smaller than minutes, so we can use them to measure short amounts of time.

T: What are other things we might measure using seconds?

S: (Discuss.)

T: Turn and tell your partner how many seconds you estimate it takes us to walk from the carpet to sit in our seats.

T: Let's use the stopwatch to measure. Go!

T: It took us ___ seconds. Use mental math to compare your estimate with the real time. How close were you? (Select a few students to share.)

T: (Display stopwatch.) The tool I'm using to measure seconds is called a **stopwatch**. We can start it and stop it to measure how much time passes by. It has two buttons. The button on the right is the start button, and the one on the left is the stop and reset button.

T: When we stopped the stopwatch, did time stop, or did we just stop measuring?

S: Time didn't stop. → We stopped measuring time by hitting the stop button. → Time keeps going. We only stopped measuring.

T: Time is **continuous**. *Continuous* means time does not stop but is always moving forward. We just use stopwatches and clocks to measure its movement.

T: Partner 1, measure and write how long it takes Partner 2 to draw a 2 by 5 array on her personal white board.

S: (Partner 1 times, and Partner 2 draws. Partner 1 writes unit form, e.g., 8 seconds.)

> **NOTES ON MULTIPLE MEANS OF ENGAGEMENT:**
>
> When introducing the stopwatch as a tool to measure time, ask students to think about where stopwatches are used in real-world contexts, for example, in swim meets and races. Then, discuss the purpose of the stopwatch in these contexts.

Student pairs take turns using a stopwatch to measure how long it takes them to do the following:

- Skip-count by fives to 60.
- Draw a 6 by 10 array.

Part 2: Explore minutes as a unit of time.

T: I look at the clock and notice that ___ minutes have passed since we walked from our tables to the carpet.

T: **Minutes** are longer than seconds. Let's find out what the length of a minute feels like. Sit quietly and measure a minute with your stopwatch. Go!

S: (Watch the stopwatch until 1 minute passes.)

T: What does a minute feel like?

S: It is *much* longer than 1 second!

T: Now, I'll time 1 minute. You turn and talk to your partner about your favorite game. Let's see if the length of 1 minute feels the same. (Time students talking.)

T: Did 1 minute feel faster or slower than when you were just watching the clock?

S: It seemed so much faster! Talking was fun!

T: How long a minute feels can change depending on what we're doing, but the measurement always stays the same. What are some other things we might use minutes to measure?

S: (Discuss.)

> **NOTES ON MULTIPLE MEANS OF ENGAGEMENT:**
>
> Possibly extend Part 1 discussion:
> T: Who was faster?
> S1: I was!
> T: Whose was neater?
> S2: Mine!
> T: In this case, was faster better?
> S: The picture was better when we went more slowly.

Student pairs take turns using a stopwatch to measure how long it takes them to do the following:

- Touch their toes and raise their hands over their heads 30 times.
- Draw 1 by 1, 2 by 2, 3 by 3, 4 by 4, and 5 by 5 arrays.

Part 3: Explore time as a continuous measurement.

T: We can use the stopwatch to start measuring how many minutes it takes to get dark outside. Will it take a long time?

S: Yes!

T: (Start the stopwatch and wait impatiently.) Should I keep measuring? (Let students react.)

T: (Stop the stopwatch.) Imagine that I measure how long it takes for all the students in this class to turn 10 years old. Is a stopwatch a good tool for measuring such a long amount of time?

S: No! It's better for measuring an amount of time that is not very long.

T: Time keeps going and going, and a stopwatch just captures a few seconds or minutes of it along the way.

> **NOTES ON MULTIPLE MEANS OF ACTION AND EXPRESSION:**
>
> When leaving the classroom for recess or lunch, consider measuring how long it takes to make a line, to go to the cafeteria, or to return to the classroom.

Lesson 1: Explore time as a continuous measurement using a stopwatch.

A STORY OF UNITS

Lesson 1 3•2

Problem Set (10 minutes)

Students should do their personal best to complete the Problem Set within the allotted 10 minutes. Some problems do not specify a method for solving. This is an intentional reduction of scaffolding that invokes MP.5, Use Appropriate Tools Strategically. Students should solve these problems using the RDW approach used for Application Problems.

For some classes, it may be appropriate to modify the assignment by specifying which problems students should work on first. With this option, let the purposeful sequencing of the Problem Set guide the selections so that problems continue to be scaffolded. Balance word problems with other problem types to ensure a range of practice. Consider assigning incomplete problems for homework or at another time during the day.

Student Debrief (10 minutes)

Lesson Objective: Explore time as a continuous measurement using a stopwatch.

The Student Debrief is intended to invite reflection and active processing of the total lesson experience.

Invite students to review their solutions for the Problem Set. They should check work by comparing answers with a partner before going over answers as a class. Look for misconceptions or misunderstandings that can be addressed in the Debrief. Guide students in a conversation to debrief the Problem Set and process the lesson.

Any combination of the questions below may be used to lead the discussion.

- Explain to your partner why the activities in Problem 5 did not take that long to complete.
- Did it take you longer to complete Problem 1 or Problem 4? Why?
- Why do we use a stopwatch?
- **Seconds** and **minutes** are units we use to measure time. How are they different?
- Does time stop when we stop measuring time with our stopwatch? Use the word *continuous* to talk about why or why not with your partner.

Lesson 1: Explore time as a continuous measurement using a stopwatch.

Exit Ticket (3 minutes)

After the Student Debrief, instruct students to complete the Exit Ticket. A review of their work will help with assessing students' understanding of the concepts that were presented in today's lesson and planning more effectively for future lessons. The questions may be read aloud to the students.

A STORY OF UNITS   Lesson 1 Problem Set 3•2

Name _____   Date _____

1. Use a stopwatch. How long does it take you to snap your fingers 10 times?

   It takes _____ to snap 10 times.

2. Use a stopwatch. How long does it take to write every whole number from 0 to 25?

   It takes _____ to write every whole number from 0 to 25.

3. Use a stopwatch. How long does it take you to name 10 animals? Record them below.

   It takes _____ to name 10 animals.

4. Use a stopwatch. How long does it take you to write 7 × 8 = 56 fifteen times? Record the time below.

   It takes _____ to write 7 × 8 = 56 fifteen times.

Lesson 1:   Explore time as a continuous measurement using a stopwatch.

A STORY OF UNITS

Lesson 1 Problem Set 3•2

5. Work with your group. Use a stopwatch to measure the time for each of the following activities.

Activity	Time
Write your full name.	_____ seconds
Do 20 jumping jacks.	
Whisper count by twos from 0 to 30.	
Draw 8 squares.	
Skip-count out loud by fours from 24 to 0.	
Say the names of your teachers from Kindergarten to Grade 3.	

6. 100 meter relay: Use a stopwatch to measure and record your team's times.

Name	Time
	Total time:

Lesson 1: Explore time as a continuous measurement using a stopwatch.

A STORY OF UNITS   **Lesson 1 Exit Ticket 3•2**

Name _____   Date _____

The table to the right shows how much time it takes each of the 5 students to do 15 jumping jacks.

Maya	16 seconds
Riley	15 seconds
Jake	14 seconds
Nicholas	15 seconds
Adeline	17 seconds

a. Who finished 15 jumping jacks the fastest?

b. Who finished their jumping jacks in the exact same amount of time?

c. How many seconds faster did Jake finish than Adeline?

A STORY OF UNITS   Lesson 1 Homework 3•2

Name _____   Date _____

1. The table to the right shows how much time it takes each of the 5 students to run 100 meters.

Samantha	19 seconds
Melanie	22 seconds
Chester	26 seconds
Dominique	18 seconds
Louie	24 seconds

   a. Who is the fastest runner?

   b. Who is the slowest runner?

   c. How many seconds faster did Samantha run than Louie?

2. List activities at home that take about the following amounts of time to complete. If you do not have a stopwatch, you can use the strategy of counting by *1 Mississippi, 2 Mississippi, 3 Mississippi, …*.

Time	Activities at home
30 seconds	Example: Tying shoelaces
45 seconds	
60 seconds	

Lesson 1: Explore time as a continuous measurement using a stopwatch.

3. Match the analog clock with the correct digital clock.

A STORY OF UNITS

Lesson 2 3•2

Lesson 2

Objective: Relate skip-counting by fives on the clock and telling time to a continuous measurement model, the number line.

Suggested Lesson Structure

- ■ Fluency Practice   (12 minutes)
- ■ Application Problem (5 minutes)
- ■ Concept Development (33 minutes)
- ■ Student Debrief   (10 minutes)
- **Total Time**   **(60 minutes)**

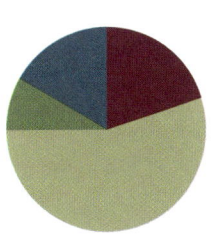

Fluency Practice (12 minutes)

- Group Counting **3.OA.1**   (3 minutes)
- Tell Time on the Clock **2.MD.7** (3 minutes)
- Minute Counting **3.MD.1**   (6 minutes)

Group Counting (3 minutes)

Note: Group counting reviews interpreting multiplication as repeated addition. Counting by sevens and eights in this activity anticipates multiplication using those units in Module 3.

Direct students to count forward and backward using the following suggested sequence, occasionally changing the direction of the count:

- Sevens to 35, emphasizing the transition from 28 to 35
- Eights to 40, emphasizing the transition from 32 to 40

Tell Time on the Clock (3 minutes)

Materials: (T) Analog clock for demonstration (S) Personal white board

Note: This activity reviews the Grade 2 standard of telling and writing time to the nearest 5 minutes. It prepares students to use the number line and clock to tell time to the nearest 5 minutes in the Concept Development.

- T: (Show an analog demonstration clock.) Start at 12 and count by 5 minutes on the clock. (Move finger from 12 to 1, 2, 3, 4, etc., as students count.)
- S: 5, 10, 15, 20, 25, 30, 35, 40, 45, 50, 55, 60.
- T: I'll show a time on the clock. Write the time on your personal white board. (Show 3:05.)
- S: (Write 3:05.)

T: (Show 2:35.)
S: (Write 2:35.)

Repeat process, varying the hour and 5-minute interval so that students read and write a variety of times to the nearest 5 minutes.

Minute Counting (6 minutes)

Note: This activity reviews the Grade 2 standard of telling and writing time to the nearest 5 minutes. It prepares students to count by 5-minute intervals on the number line and clock in the Concept Development. Students also practice group counting strategies for multiplication in the context of time.

Use the process outlined for this activity in Lesson 1. Direct students to count by 5 minutes to the hour, the half hour, and the quarter hour. Repeat the process using the following suggested sequence:

- 6 minutes, counting to the half hour and hour
- 3 minutes, counting to a quarter past the hour and half hour
- 10 minutes, counting up to 1 hour
- 9 minutes, counting to 45 and emphasizing the transition from 36 to 45

Application Problem (5 minutes)

Christine has 12 math problems for homework. It takes her 5 minutes to complete each problem. How many minutes does it take Christine to finish all 12 problems?

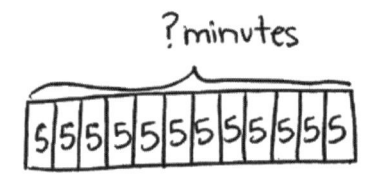

? minutes

| 5 | 5 | 5 | 5 | 5 | 5 | 5 | 5 | 5 | 5 | 5 | 5 |

12 × 5 minutes = 60 minutes
It takes Christine 60 minutes to finish her homework.

Note: This problem anticipates the Concept Development. It activates prior knowledge from Grade 2 about math with minutes. Twelve is a new factor. If students are unsure about how to multiply 12 groups of 5, encourage them to solve by skip-counting. They can also use the distributive property, 10 fives + 2 fives or 6 fives + 6 fives. Students use the solution to this problem as a springboard for modeling 12 intervals of 5 minutes on the number line in the Concept Development.

| A STORY OF UNITS | Lesson 2 | 3•2 |

Concept Development (33 minutes)

Materials: (T) Analog clock for demonstration (S) Personal white board, tape diagram (Template 1), two clocks (Template 2), centimeter ruler

Part 1: Draw a number line and relate skip-counting by fives to skip-counting intervals of 5 minutes.

Students place the tape diagram template in personal white boards.

T: Model the Application Problem using the tape diagram on the template.

S: (Model.)

Template 1

Guide discussion so that students articulate the following: the tape diagram is divided into 12 parts, with each part representing the time it takes Christine to do one math problem; the whole tape diagram represents a total of 60 minutes.

T: A different way to model this problem is to use a number line. Let's use our tape diagram to help us draw a number line that represents a total of 60 minutes.

T: Draw a line a few centimeters below the tape diagram. Make it the same length as the tape diagram. Make tick marks on the number line where units are divided on the tape diagram. (Model each step as students follow along.)

MP.4

T: What do you notice about the relationship between the tape diagram and the number line?

S: The lines are in the same place. → They have the same number of parts.

T: What part of the tape diagram do the spaces between tick marks represent?

S: The units. → The time it takes to do each math problem. → They each represent 5 minutes.

T: We know from yesterday that time doesn't stop. It was happening before Christine started her homework, and it keeps going after she's finished. To show that time is continuous, we'll extend our number line on both sides and add arrows to it. (Model.)

S: (Extend number lines and add arrows.)

T: Let's label our number lines. The space between 2 tick marks represents a 5-minute **interval**. Write 0 under the first tick mark on the left. Then, skip-count by fives. As you count, write each number under the next tick mark. Stop when you've labeled 60. (Model as students follow along.)

T: The space between 2 marks represents one 5-minute interval. How many minutes are in the interval from 0 to 10? From 0 to 60? From 15 to 30?

S: From 0 to 10 is 10 minutes, from 0 to 60 is 60 minutes, and from 15 to 30 is 15 minutes.

T: Let's use the number line to find how many minutes it takes Christine to do 4 math problems. (Place finger at 0. Move to 5, 10, 15, and 20 as you count 1 problem, 2 problems, 3 problems, 4 problems.) It takes Christine 20 minutes to do 4 math problems. Use the word *interval* to explain to your partner how we used the number line to figure that out.

S: (Discuss.)

Use guided practice to find how long it takes Christine to solve 7, 9, and 11 problems.

A STORY OF UNITS   Lesson 2 3•2

Part 2: Use a number line to tell time to the nearest 5 minutes within 1 hour.

T: Use your ruler to draw a 12-centimeter number line. (Model as students follow along.)
T: How many 5-minute intervals will the number line need to represent a total of 60 minutes?
S: Twelve!
T: Marking 12 equally spaced intervals is difficult! How can the ruler help do that?
S: It has 12 centimeters. → The centimeters show us where to draw tick marks.
T: Use the centimeters on your ruler to draw tick marks for the number line. (Model.)
S: (Use rulers to draw tick marks.)
T: Just like on the first number line, we'll need to show that time is continuous. Extend each side of your number line and make arrows. Then skip-count to label each 5-minute interval starting with 0 and ending with 60. (Model while students follow along.)

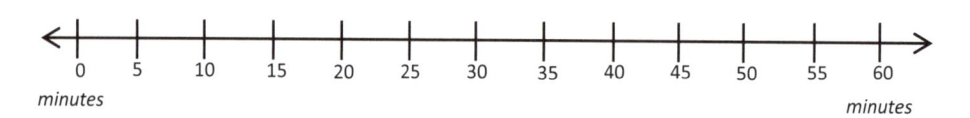

```
←——+——+——+——+——+——+——+——+——+——+——+——+——→
   0   5  10  15  20  25  30  35  40  45  50  55  60
minutes                                                minutes
```

T: How many minutes are labeled on our number line?
S: 60 minutes.
T: There are 60 minutes between 1:00 p.m. and 2:00 p.m. Let's use the number line to model exactly when we will do the activities on our class schedule that happen between 1:00 p.m. and 2:00 p.m.
T: Below the 0 tick mark, write 1:00 p.m. Below the 60 tick mark, write 2:00 p.m. (Model.)
S: (Label as shown below.)

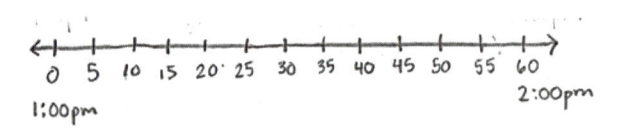

```
←——+——+——+——+——+——+——+——+——+——+——+——+——→
   0   5  10  15  20  25  30  35  40  45  50  55  60
1:00pm                                              2:00pm
```

T: Now this number line shows the hour between 1:00 p.m. and 2:00 p.m.
T: We start recess at 1:10 p.m. Is that time between 1:00 p.m. and 2:00 p.m.?
S: (Agree.)
T: To find that spot on the number line, I'll put my finger on 1:00 and move it to the right as I skip-count intervals until I reach 1:10. Remind me, what are we counting by?
S: Fives!
T: (Model, with students chorally counting along.)

> **NOTES ON MULTIPLE MEANS OF ACTION AND EXPRESSION:**
>
> You need not use 1 p.m. to 2 p.m. as the interval; pick an hour that is relevant to today's class. As students determine the number of 5-minute intervals on the number line, some may count tick marks instead of spaces and get an answer of 13. Watch for this misconception and guide students to make a distinction between tick marks and intervals if necessary.

Lesson 2: Relate skip-counting by fives on the clock and telling time to a continuous measurement model, the number line.

A STORY OF UNITS

Lesson 2 3•2

T: I'll draw a dot on the spot where the tick mark and number line cross and label it *R* for recess. (Draw and label as shown on the right.) That dot shows the location of a **point**. Finding and drawing a point is called **plotting** a position on the number line.

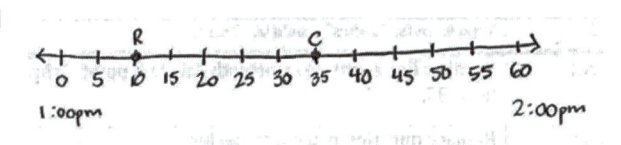

T: At 1:35 p.m., we'll start science. Is 1:35 p.m. between 1:00 p.m. and 2:00 p.m.?

S: (Agree.)

T: Plot 1:35 p.m. as a point on your number line. Label it C.

S: (Plot a point on the number line at 1:35.)

Continue guided practice using the following suggested sequence: 1:45 p.m. and 2:00 p.m.

T: How does the number line you've labeled compare to the analog clock on the wall?

S: We count the minutes by fives on both. → The clock is like the number line wrapped in a circle.

Part 3: Relate the number line to the clock and tell time to the nearest 5 minutes.

Students have Template 2 (two clocks) ready. Display a clock face without hands.

T: We counted by fives to plot minutes on a number line, and we'll do the same on a clock.

T: How many 5-minute intervals show 15 minutes on a clock?

S: 3 intervals.

T: We started at 0 on the number line, but a clock has no 0. Where is the starting point on a clock?

S: The 12.

T: Let's count each 5-minute interval and plot a point on the clock to show 15 minutes. (Model.)

Options for further practice:

- Plot 30 minutes, 45 minutes, and 55 minutes using the process above.
- Write 9:15 a.m., 3:30 p.m., and 7:50 a.m. on the board as they would appear on a digital clock, or say the time rather than write it. Students copy each time, plot points, and draw hands to show that time. (Model drawing hands with 10:20 a.m.)

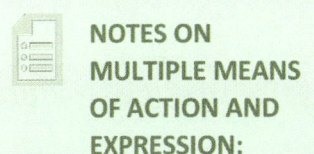

> **NOTES ON MULTIPLE MEANS OF ACTION AND EXPRESSION:**
>
> Extend the discussion by inviting students to discuss whether or not 12:55 p.m. and 2:15 p.m. can be plotted on this number line. Help them reason about their answer and think about where the times might be plotted, given the continuity of time.

> **NOTES ON MULTIPLE MEANS OF REPRESENTATION:**
>
> Activate prior knowledge about the minute hand and hour hand learned in Grade 2 Module 2. Review their difference in purpose, as well as in length.

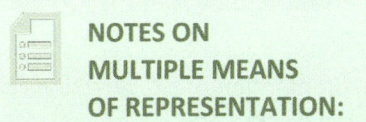

Template 2

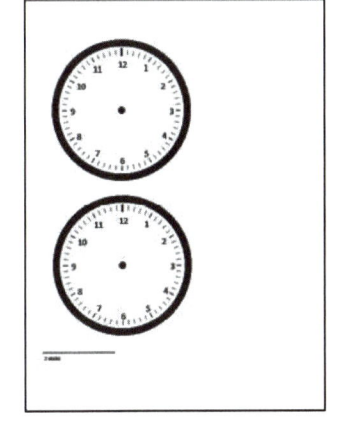

Lesson 2: Relate skip-counting by fives on the clock and telling time to a continuous measurement model, the number line.

A STORY OF UNITS   Lesson 2   3•2

Problem Set (10 minutes)

Students should do their personal best to complete the Problem Set within the allotted 10 minutes. For some classes, it may be appropriate to modify the assignment by specifying which problems they work on first. Some problems do not specify a method for solving. Students should solve these problems using the RDW approach used for Application Problems.

Student Debrief (10 minutes)

Lesson Objective: Relate skip-counting by fives on the clock and telling time to a continuous measurement model, the number line.

The Student Debrief is intended to invite reflection and active processing of the total lesson experience.

Invite students to review their solutions for the Problem Set. They should check work by comparing answers with a partner before going over answers as a class. Look for misconceptions or misunderstandings that can be addressed in the Debrief. Guide students in a conversation to debrief the Problem Set and process the lesson.

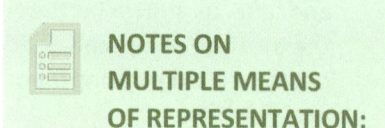

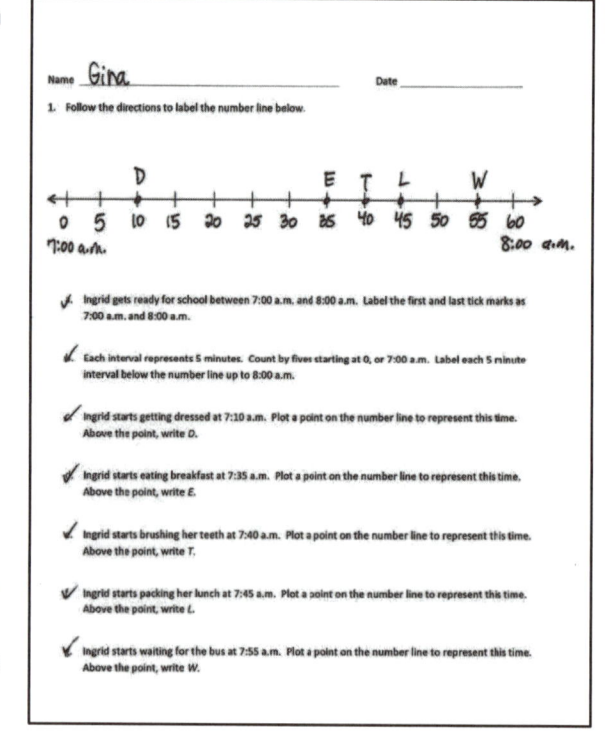

Any combination of the questions below may be used to lead the discussion.

- In Problem 2, what information was important for **plotting** the **point** on the number line that matched the time shown on each clock?
- Each **interval** on the analog clock is labeled with the numbers 1–12. Compare those with our labels from 0 to 60 on the number line. What do the labels represent on both tools?
- How does multiplication using units of 5 help you read or measure time?
- Students may have different answers for Problem 4 (11:25 p.m. may come before or after 11:20 a.m.). Allow students with either answer a chance to explain their thinking.
- How did our minute counting and time telling activities in today's Fluency Practice help you with the rest of the lesson?
- Look at the number line used for Problem 2. Where do you think 5:38 would be? (This anticipates Lesson 3 by counting by fives and then ones on a number line.)

> **NOTES ON MULTIPLE MEANS OF REPRESENTATION:**
>
> Problem 4 is likely to pose the biggest challenge. It requires understanding the difference between a.m. and p.m. This concept was introduced in Grade 2. One option would be to review it with students before they begin the Problem Set. Another option would be to allow them to grapple with the question and support understanding through the Student Debrief.

Lesson 2: Relate skip-counting by fives on the clock and telling time to a continuous measurement model, the number line.

Lesson 2 3•2

Exit Ticket (3 minutes)

After the Student Debrief, instruct students to complete the Exit Ticket. A review of their work will help with assessing students' understanding of the concepts that were presented in today's lesson and planning more effectively for future lessons. The questions may be read aloud to the students

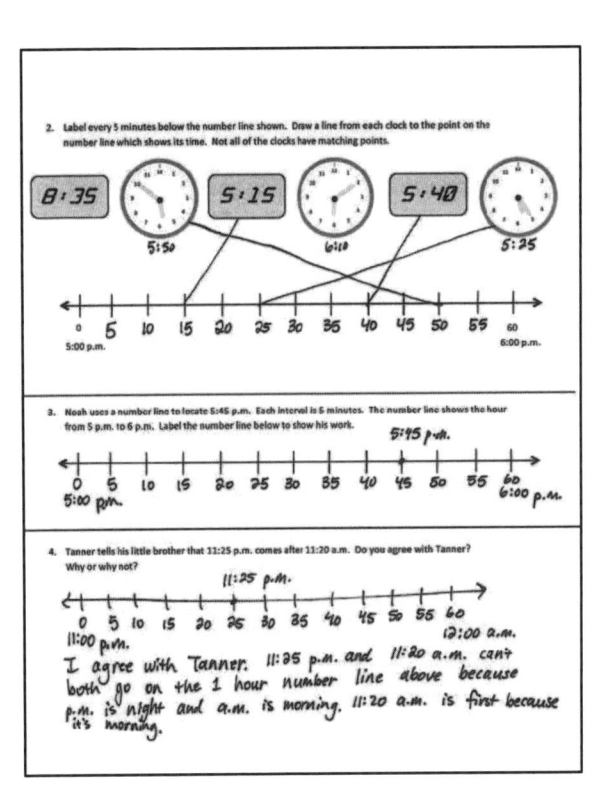

Lesson 2: Relate skip-counting by fives on the clock and telling time to a continuous measurement model, the number line.

Name _____ Date _____

1. Follow the directions to label the number line below.

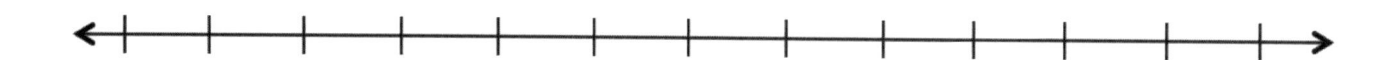

   ←—+——+——+——+——+——+——+——+——+——+——+——+—→

   a. Ingrid gets ready for school between 7:00 a.m. and 8:00 a.m. Label the first and last tick marks as 7:00 a.m. and 8:00 a.m.

   b. Each interval represents 5 minutes. Count by fives starting at 0, or 7:00 a.m. Label each 5-minute interval below the number line up to 8:00 a.m.

   c. Ingrid starts getting dressed at 7:10 a.m. Plot a point on the number line to represent this time. Above the point, write *D*.

   d. Ingrid starts eating breakfast at 7:35 a.m. Plot a point on the number line to represent this time. Above the point, write *E*.

   e. Ingrid starts brushing her teeth at 7:40 a.m. Plot a point on the number line to represent this time. Above the point, write *T*.

   f. Ingrid starts packing her lunch at 7:45 a.m. Plot a point on the number line to represent this time. Above the point, write *L*.

   g. Ingrid starts waiting for the bus at 7:55 a.m. Plot a point on the number line to represent this time. Above the point, write *W*.

2. Label every 5 minutes below the number line shown. Draw a line from each clock to the point on the number line which shows its time. Not all of the clocks have matching points.

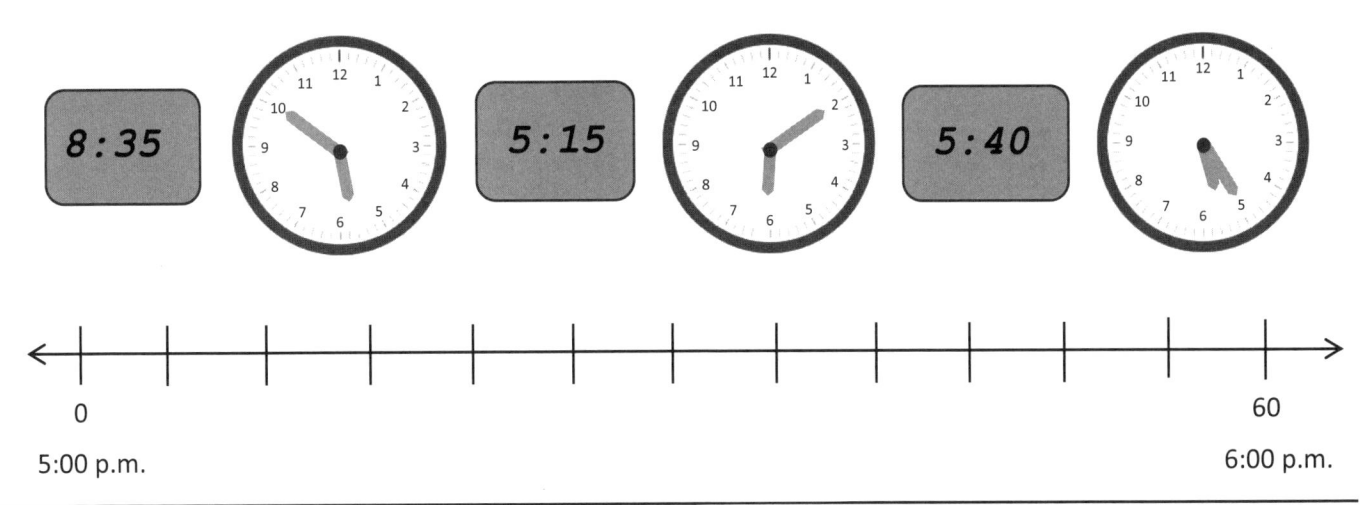

8:35   5:15   5:40

0   60
5:00 p.m.   6:00 p.m.

3. Noah uses a number line to locate 5:45 p.m. Each interval is 5 minutes. The number line shows the hour from 5 p.m. to 6 p.m. Label the number line below to show his work.

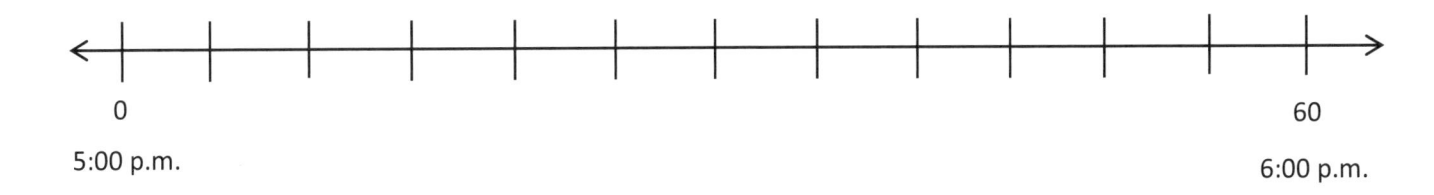

0   60
5:00 p.m.   6:00 p.m.

4. Tanner tells his little brother that 11:25 p.m. comes after 11:20 a.m. Do you agree with Tanner? Why or why not?

Name _____ Date _____

The number line below shows a math class that begins at 10:00 a.m. and ends at 11:00 a.m. Use the number line to answer the following questions.

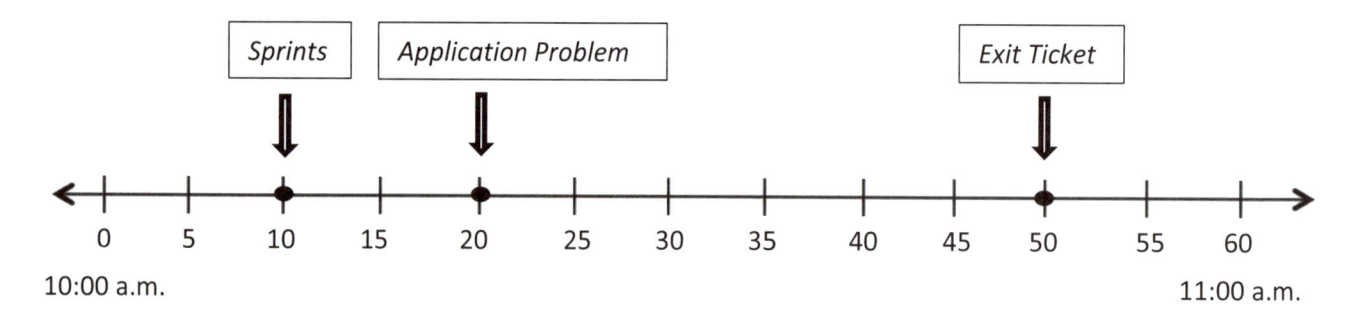

a. What time do Sprints begin?

b. What time do students begin the Application Problem?

c. What time do students work on the Exit Ticket?

d. How long is math class?

Name _____ Date _____

Follow the directions to label the number line below.

←|——|——|——|——|——|——|——|——|——|——|——|——|→

a. The basketball team practices between 4:00 p.m. and 5:00 p.m. Label the first and last tick marks as 4:00 p.m. and 5:00 p.m.

b. Each interval represents 5 minutes. Count by fives starting at 0, or 4:00 p.m. Label each 5-minute interval below the number line up to 5:00 p.m.

c. The team warms up at 4:05 p.m. Plot a point on the number line to represent this time. Above the point, write *W*.

d. The team shoots free throws at 4:15 p.m. Plot a point on the number line to represent this time. Above the point, write *F*.

e. The team plays a practice game at 4:25 p.m. Plot a point on the number line to represent this time. Above the point, write *G*.

f. The team has a water break at 4:50 p.m. Plot a point on the number line to represent this time. Above the point, write *B*.

g. The team reviews their plays at 4:55 p.m. Plot a point on the number line to represent this time. Above the point, write *P*.

tape diagram

A STORY OF UNITS Lesson 2 Template 2 3•2

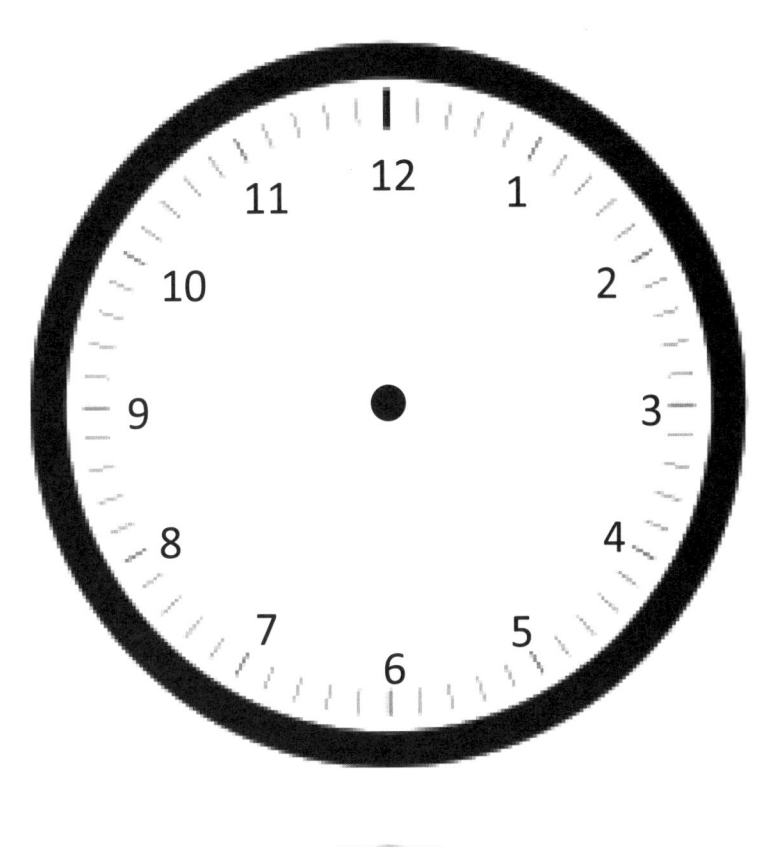

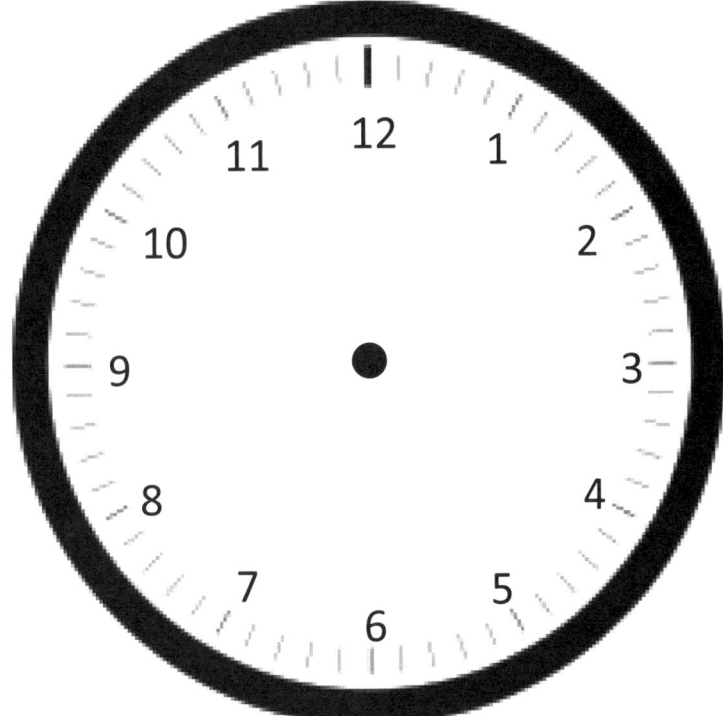

two clocks

Lesson 2: Relate skip-counting by fives on the clock and telling time to a continuous measurement model, the number line.

A STORY OF UNITS　　　　　　　　　　　　　　　　　　　　　　　　　　　　　Lesson 3　3•2

Lesson 3

Objective: Count by fives and ones on the number line as a strategy to tell time to the nearest minute on the clock.

Suggested Lesson Structure

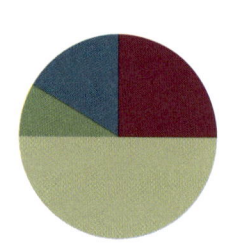

■ Fluency Practice　　　　(15 minutes)
■ Application Problem　　 (5 minutes)
■ Concept Development　 (30 minutes)
■ Student Debrief　　　　 (10 minutes)
　Total Time　　　　　　**(60 minutes)**

Fluency Practice (15 minutes)

- Tell Time on the Clock **2.MD.7**　　(3 minutes)
- Decompose 60 Minutes **3.MD.1**　　(6 minutes)
- Minute Counting **3.MD.1**　　　　　 (3 minutes)
- Group Counting **3.OA.1**　　　　　　(3 minutes)

Tell Time on the Clock (3 minutes)

Materials: (T) Analog clock for demonstration (S) Personal white board

Note: This activity reviews the Grade 2 standard of telling and writing time to the nearest 5 minutes. It reviews Lesson 2 and prepares students to count by 5 minutes and some ones in this lesson.

　T: (Show an analog demonstration clock.) Start at 12 and count by 5 minutes on the clock. (Move finger from 12 to 1, 2, 3, 4, etc., as students count.)
　S: 5, 10, 15, 20, 25, 30, 35, 40, 45, 50, 55, 60.
　T: I'll show a time on the clock. Write the time on your personal white board. (Show 4:00.)
　S: (Write 4:00.)
　T: (Show 4:15.)
　S: (Write 4:15.)

Repeat process, varying the hour and 5-minute interval so that students read and write a variety of times to the nearest 5 minutes.

Decompose 60 Minutes (6 minutes)

Materials: (S) Personal white board

Note: Decomposing 60 minutes using a number bond helps students relate part–whole thinking to telling time.

- T: (Project a number bond with 60 minutes written as the whole.) There are 60 minutes in 1 hour.
- T: (Write 50 minutes as one of the parts.) On your board, draw this number bond and complete the unknown part.
- S: (Draw number bond with 10 minutes, completing the unknown part.)

Repeat the process for 30 minutes, 40 minutes, 45 minutes, and 35 minutes.

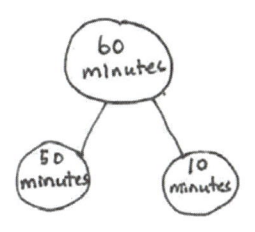

Minute Counting (3 minutes)

Note: Students practice counting strategies for multiplication in the context of time. This activity prepares students for telling time to the nearest minute and builds skills for using mental math to add and subtract minute intervals in Lesson 5.

Use the process outlined for this activity in Lesson 1. Direct students to count by 5 minutes to an hour, half hour, and quarter hour.

- 6 minutes, counting to 1 hour, and naming half hour and 1 hour intervals as such
- 3 minutes, counting to 30 minutes, and naming the quarter hour and half hour intervals as such
- 9 minutes, counting to quarter 'til 1 hour
- 10 minutes, using the following sequence: 10 minutes, 20 minutes, half hour, 40 minutes, 50 minutes, 1 hour

> **NOTES ON MULTIPLE MEANS OF REPRESENTATION:**
>
> The vocabulary *half-past*, *quarter-past* and *quarter 'til* might be challenging for students. Consider reviewing the phrases by writing the words on the board and having students read them chorally. As you define the phrases for students, next to each one, draw a circle with clock hands pointing to the place that corresponds to the language. Leave it on the board for students to reference during this activity.

Group Counting (3 minutes)

Note: Group counting reviews the interpretation of multiplication as repeated addition. Counting by sevens, eights, and nines in this activity anticipates multiplication using those units in Module 3.

Direct students to count forward and backward using the following suggested sequence, occasionally changing the direction of the count:

- Sevens to 42, emphasizing the transition from 35 to 42
- Eights to 48, emphasizing the transition from 40 to 48
- Nines to 54, emphasizing the transition from 45 to 54

Lesson 3: Count by fives and ones on the number line as a strategy to tell time to the nearest minute on the clock.

A STORY OF UNITS   Lesson 3 3•2

Application Problem (5 minutes)

There are 12 tables in the cafeteria. Five students sit at each of the first 11 tables. Three students sit at the last table. How many students are sitting at the 12 tables in the cafeteria?

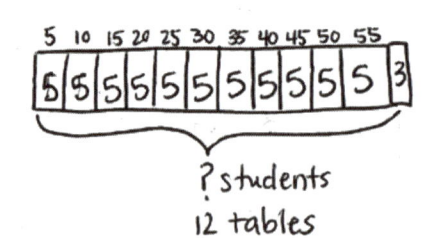

```
   5 10 15 20 25 30 35 40 45 50 55
  ┌─┬─┬─┬─┬─┬─┬─┬─┬─┬─┬─┬─┐
  │5│5│5│5│5│5│5│5│5│5│5│3│
  └─┴─┴─┴─┴─┴─┴─┴─┴─┴─┴─┴─┘
       ? students
       12 tables
```

55 + 3 = 58

There are 58 students sitting in the cafeteria.

Note: This problem activates prior knowledge from Module 1 about multiplying by 5. Students relate this work from the Application Problem to modeling minutes on the number line in the Concept Development.

Concept Development (30 minutes)

Materials: (T) Analog clock for demonstration (S) Personal white board, centimeter ruler, clock (Template) (pictured to the right)

Template

Problem 1: Count minutes by fives and ones on a number line.

T: Use your ruler to draw a 12-centimeter line on your personal white board. Start at the 0 mark, and make a tick mark at each centimeter up to the number 12. Label the first tick mark 0 and the last tick mark 60. Then, count by fives from 0 to 60 to label each interval, like we did in the last lesson.

S: (Draw and label a number line as shown.)

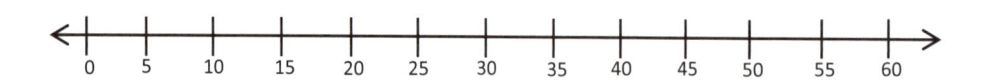

```
←─┬──┬──┬──┬──┬──┬──┬──┬──┬──┬──┬──┬──→
  0  5  10 15 20 25 30 35 40 45 50 55 60
```

T: Put your finger on 0. Count by ones from 0 to 5. What numbers did you count between 0 and 5?

S: 1, 2, 3, and 4.

T: We could draw tick marks, but let's instead imagine they are there. Can you see them?

S: Yes!

T: Put your finger on 5. Count on by ones from 5 to 10. What numbers did you count between 5 and 10?

S: 6, 7, 8, and 9.

T: Can you imagine those tick marks, too?

> **NOTES ON MULTIPLE MEANS OF ACTION AND EXPRESSION:**
>
> Use preprinted number lines for students with fine motor or perception difficulties. You can also have students actually draw all the tick marks, but be aware this may encourage counting all when the objective is to count by fives and ones.

A STORY OF UNITS   Lesson 3 3•2

S: Yes!
T: Let's find 58 minutes on the number line. Put your finger on 0. Count by five to 55.
S: (Count 11 fives.)
T: Let's draw the tick marks from 55 to 60. Count with me as I draw the tick marks from 55 to 60. Start at 55, which is already there.
S: 55, (begin drawing) 56, 57, 58, 59, (stop drawing) 60.
T: How many ticks did I draw?
S: 4.
T: Go ahead and draw yours. (Allow students time to draw.)
T: Count on by ones to find 58 using the tick marks we made in the interval between 55 and 60.

MP.6

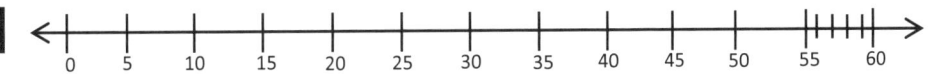

S: (Count on by ones and say numbers aloud.) 56, 57, 58.
T: How many fives did we count?
S: 11.
T: How many ones did we count?
S: 3.
T: 11 fives + 3. How can we write that as multiplication? Discuss with your partner.
S: (11 × 5) + 3.
T: Discuss with a partner how our modeling with the number line relates to the Application Problem.
S: (Discuss.)

Repeat the process with other combinations of fives and ones, such as (4 × 5) + 2 and (0 × 5) + 4.

T: Which units did we count by on the number line to solve these problems?
S: Fives and ones.
T: Whisper to your partner. What steps did we take to solve these problems on the number line?
S: (Discuss.)

Problem 2: Count by fives and ones on a number line to tell time to the nearest minute.

T: I arrived at school this morning at 7:37 a.m. Let's find that time on our number line. Label 7:00 a.m. above the 0 mark and 8:00 a.m. above the 60 mark.
S: (Label 7:00 a.m. and 8:00 a.m.)
T: Which units should we count by to get to 7:37?
S: Count by fives to 7:35 and then by ones to 7:37.
T: How many fives?
S: 7 fives.
T: How many ones?
S: 2 ones.

Lesson 3: Count by fives and ones on the number line as a strategy to tell time to the nearest minute on the clock.

A STORY OF UNITS   Lesson 3 3•2

T: Let's move our fingers over 7 fives and 2 ones on the number line.
S: (Move fingers and count.)
T: Give me a number sentence.
S: (7 × 5) + 2 = 37.
T: Plot the point on your number line.

Repeat the process with other times that can be plotted on this same number line, such as 7:13 a.m., 7:49 a.m., and 7:02 a.m.

Problem 3: Count by fives and ones on a clock to tell time to the nearest minute.

T: Insert the clock template in your personal white board. How is the clock similar to our number line?
S: There are 4 tick marks between the numbers on both. → They both have intervals of 5 with 4 marks in between.
T: What do the small tick marks represent on the clock?
S: Ones. → 1 minute!
T: We can use a clock just like we use a number line to tell time because a clock is a circular number line. Imagine twisting our number line into a circle. In your mind's eye, at what number do the ends of your number line connect?
S: At the 12.
T: The 12 on the clock represents the end of one hour and the beginning of another.
T: (Project the analog clock and draw the hands as shown.) This clock shows what time I woke up this morning. Draw the minute hand on your clock to look like mine.
S: (Draw the hand on the clock template.)
T: Let's find the minutes by counting by fives and ones. Put your finger on the 12—the zero—and count by fives with me.
S: (Move finger along the clock and count by fives to 45.)
T: (Stop at 45.) How many minutes?
S: 45.
T: Let's count on by ones until we get to the minute hand. Move your finger and count on with me.
S: 46, 47, 48. (Move finger and count on by ones.)
T: How many minutes?
S: 48.
T: Draw the hour hand. How many hours?
S: 5.
T: What is the time?
S: 5:48 a.m.
T: Write the time on your personal white boards.
S: (Write 5:48 a.m.)

Lesson 3: Count by fives and ones on the number line as a strategy to tell time to the nearest minute on the clock.

| A STORY OF UNITS | Lesson 3 | 3•2 |

Repeat the process of telling time to the nearest minute, providing a small context for each example. Use the following suggested sequence: 12:14 a.m. and 2:28 p.m.

> T: Can anyone share another strategy they used to tell the time on the clock for 2:28 p.m. other than counting by fives and ones from the 0 minute mark?
>
> S: I started at 2:30 p.m. and counted back 2 minutes to get to 2:28 p.m.

Problem Set (10 minutes)

Students should do their personal best to complete the Problem Set within the allotted 10 minutes. For some classes, it may be appropriate to modify the assignment by specifying which problems they work on first. Some problems do not specify a method for solving. Students should solve these problems using the RDW approach used for Application Problems.

Student Debrief (10 minutes)

Lesson Objective: Count by fives and ones on the number line as a strategy to tell time to the nearest minute on the clock.

The Student Debrief is intended to invite reflection and active processing of the total lesson experience.

Invite students to review their solutions for the Problem Set. They should check work by comparing answers with a partner before going over answers as a class. Look for misconceptions or misunderstandings that can be addressed in the Debrief. Guide students in a conversation to debrief the Problem Set and process the lesson.

Any combination of the questions below may be used to lead the discussion.

- Look at Problem 1. Talk to a partner: How is the number line similar to the analog clock? How is it different?
- What strategy did you use to draw the hands on the clock in Problem 3?
- Look at Problem 4. How many fives did you count by? Write a multiplication equation to show that. How many ones did you count on by? Write a multiplication equation to show that. How many minutes altogether?
- How does the tape diagram that many of us drew to solve the Application Problem relate to the first number line we drew in the Concept Development?
- Look at Problem 5. Can you share another strategy you used to tell the time on the clock other than counting by fives and ones from the 0 minute mark?

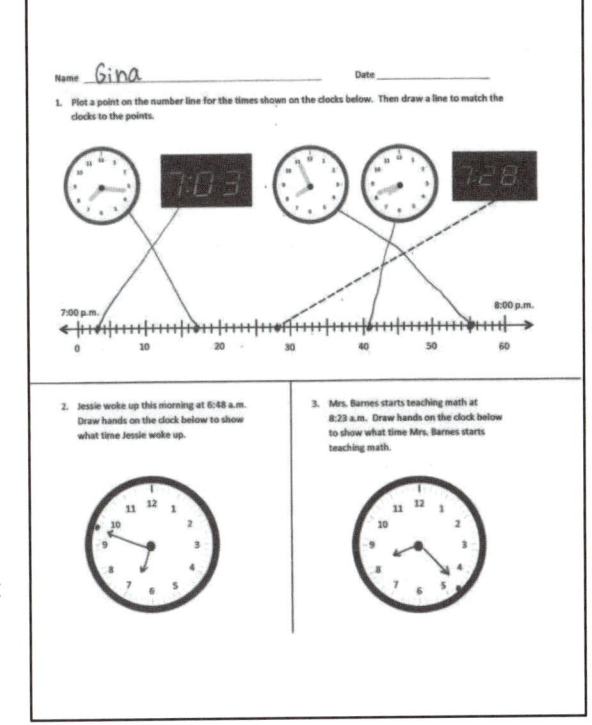

A STORY OF UNITS   Lesson 3   3•2

- (In anticipation of Lesson 4, which involves solving word problems with time intervals, have students discuss Problem 5(b).) How is Problem 5(b) different from the rest of the problems? How can you solve Problem 5(b)?

Exit Ticket (3 minutes)

After the Student Debrief, instruct students to complete the Exit Ticket. A review of their work will help with assessing students' understanding of the concepts that were presented in today's lesson and planning more effectively for future lessons. The questions may be read aloud to the students.

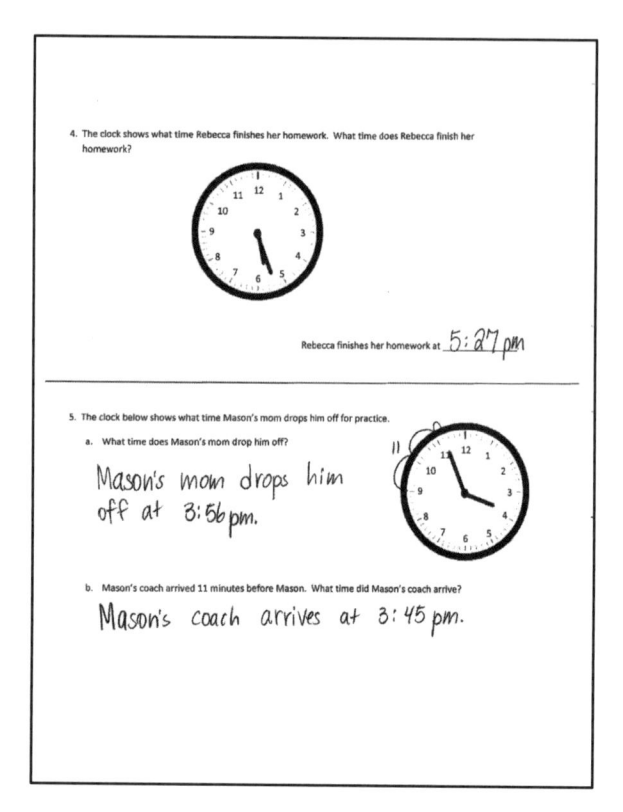

---

4. The clock shows what time Rebecca finishes her homework. What time does Rebecca finish her homework?

Rebecca finishes her homework at **5:27 pm**

5. The clock below shows what time Mason's mom drops him off for practice.

   a. What time does Mason's mom drop him off?

   Mason's mom drops him off at 3:56 pm.

   b. Mason's coach arrived 11 minutes before Mason. What time did Mason's coach arrive?

   Mason's coach arrives at 3:45 pm.

A STORY OF UNITS

Lesson 3 Problem Set 3•2

Name _____ Date _____

1. Plot a point on the number line for the times shown on the clocks below. Then, draw a line to match the clocks to the points.

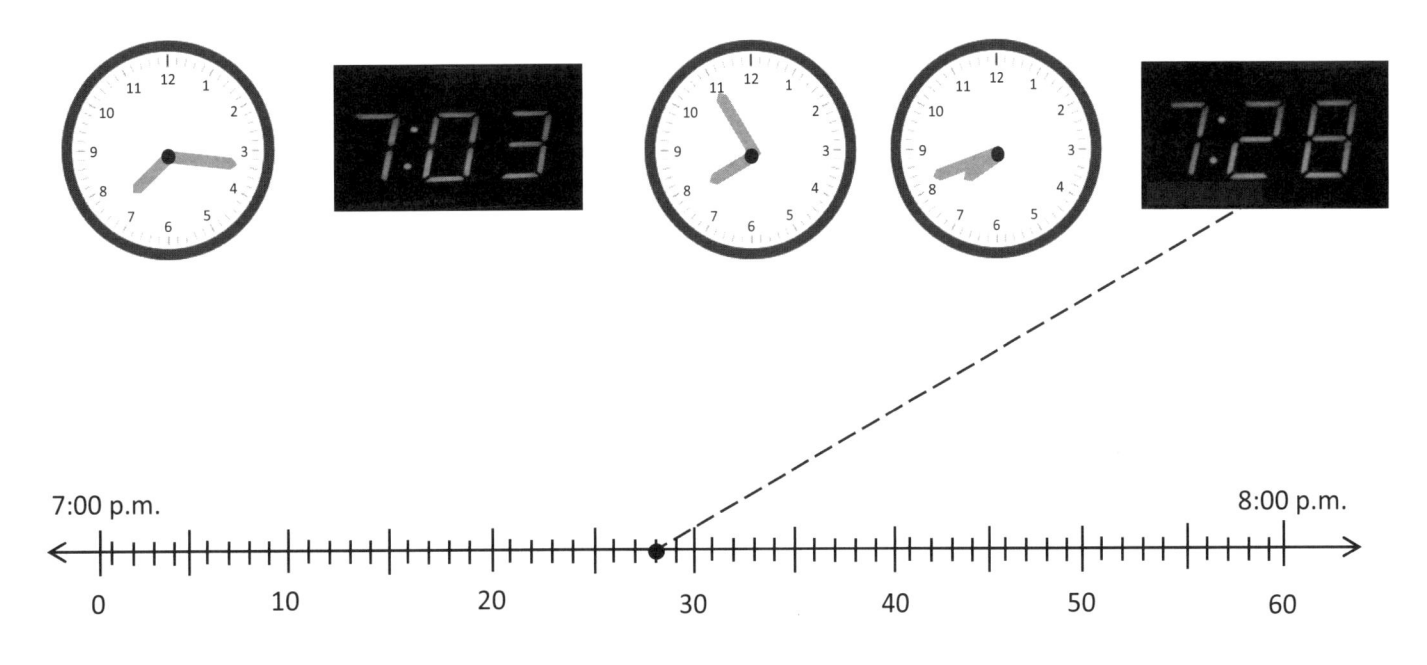

7:00 p.m. ←|||||||||||||||||||||||||||||||||||||||||||||||||||||||||||||→ 8:00 p.m.
0   10   20   30   40   50   60

2. Jessie woke up this morning at 6:48 a.m. Draw hands on the clock below to show what time Jessie woke up.

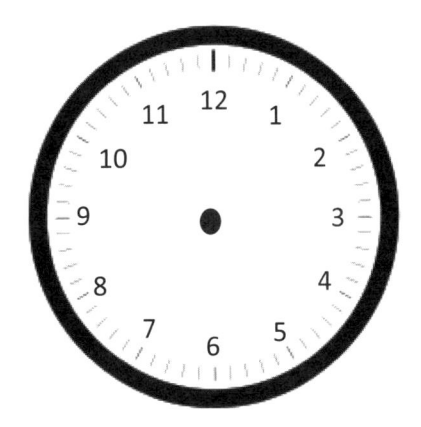

3. Mrs. Barnes starts teaching math at 8:23 a.m. Draw hands on the clock below to show what time Mrs. Barnes starts teaching math.

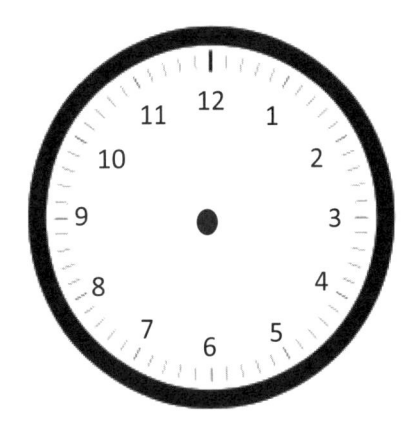

Lesson 3: Count by fives and ones on the number line as a strategy to tell time to the nearest minute on the clock.

EUREKA MATH

43

This work is derived from Eureka Math™ and licensed by Great Minds. ©2015 Great Minds. eureka-math.org

4. The clock shows what time Rebecca finishes her homework. What time does Rebecca finish her homework?

Rebecca finishes her homework at _____.

5. The clock below shows what time Mason's mom drops him off for practice.

   a. What time does Mason's mom drop him off?

   b. Mason's coach arrived 11 minutes before Mason. What time did Mason's coach arrive?

A STORY OF UNITS　　　　　　　　　　　　　　　　　　　　Lesson 3 Exit Ticket 3•2

Name _____   Date _____

The clock shows what time Jason gets to school in the morning.

Arrival at School

a. What time does Jason get to school?

b. The first bell rings at 8:23 a.m. Draw hands on the clock to show when the first bell rings.

First Bell Rings

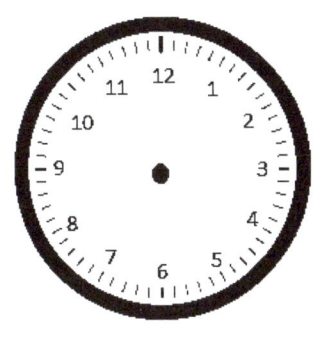

c. Label the first and last tick marks 8:00 a.m. and 9:00 a.m. Plot a point to show when Jason arrives at school. Label it *A*. Plot a point on the line when the first bell rings and label it *B*.

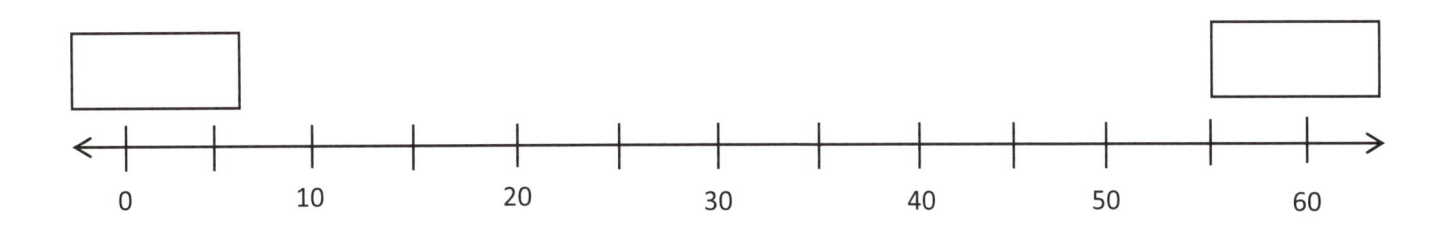

0   10   20   30   40   50   60

Lesson 3:   Count by fives and ones on the number line as a strategy to tell time to the nearest minute on the clock.

45

A STORY OF UNITS   Lesson 3 Homework 3•2

Name _____   Date _____

1. Plot points on the number line for each time shown on a clock below. Then, draw lines to match the clocks to the points.

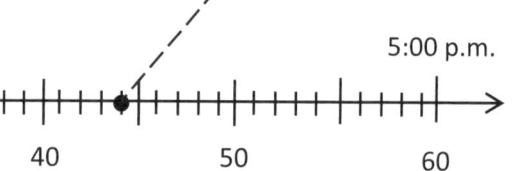

4:00 p.m.   5:00 p.m.

0   10   20   30   40   50   60

---

2. Julie eats dinner at 6:07 p.m. Draw hands on the clock below to show what time Julie eats dinner.

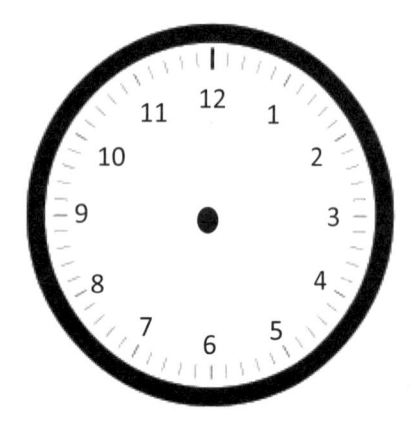

3. P.E. starts at 1:32 p.m. Draw hands on the clock below to show what time P.E. starts.

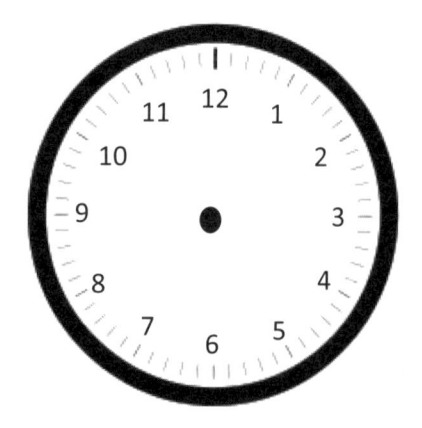

46   Lesson 3:   Count by fives and ones on the number line as a strategy to tell time to the nearest minute on the clock.

EUREKA MATH

This work is derived from Eureka Math ™ and licensed by Great Minds. ©2015 Great Minds. eureka-math.org

4. The clock shows what time Zachary starts playing with his action figures.

   a. What time does he start playing with his action figures?

   b. He plays with his action figures for 23 minutes. What time does he finish playing?

   c. Draw hands on the clock to the right to show what time Zachary finishes playing.

   d. Label the first and last tick marks with 2:00 p.m. and 3:00 p.m. Then, plot Zachary's start and finish times. Label his start time with a *B* and his finish time with an *F*.

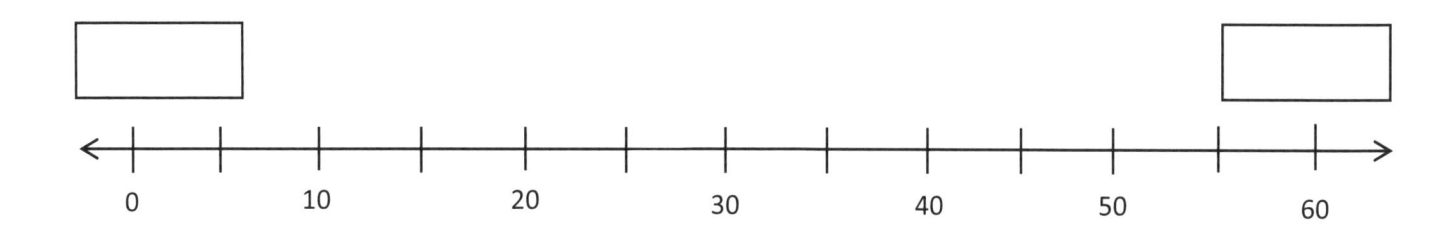

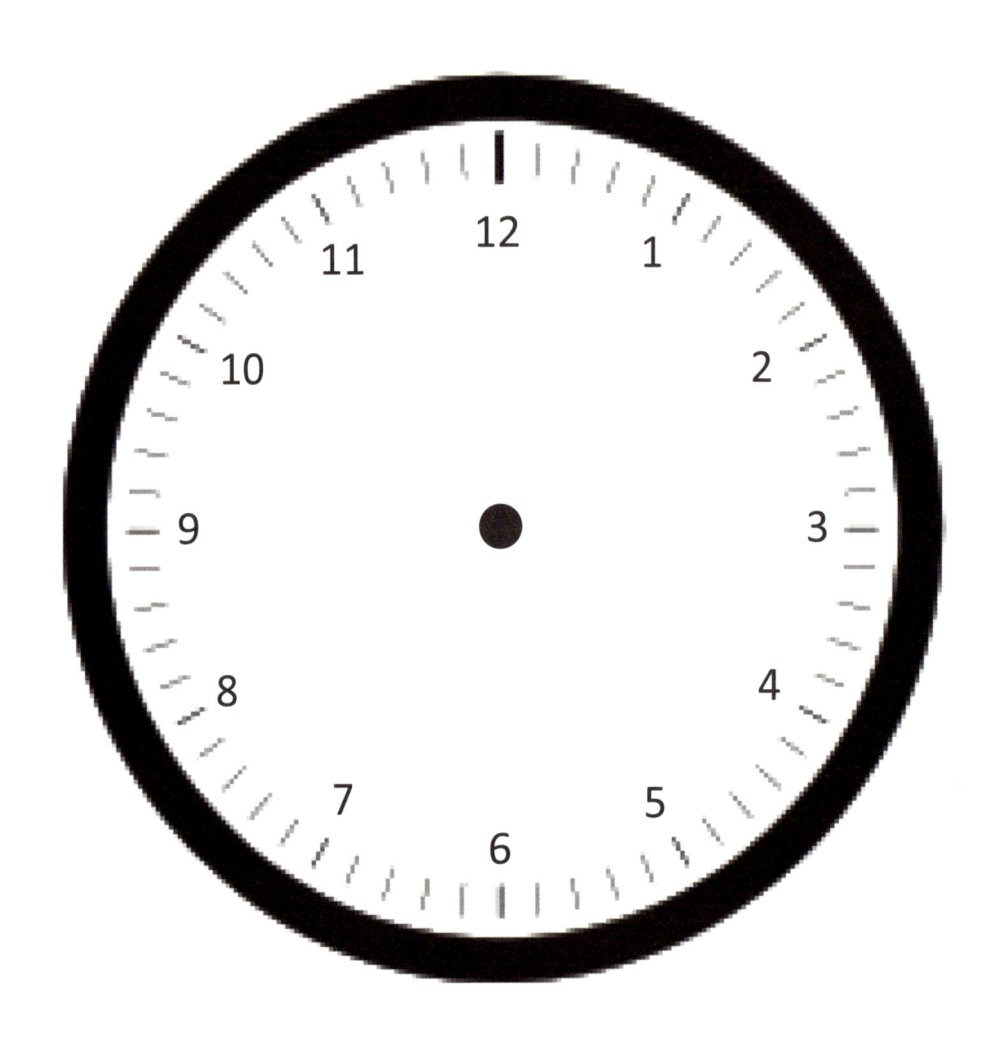

clock

Lesson 4

Objective: Solve word problems involving time intervals within 1 hour by counting backward and forward using the number line and clock.

Suggested Lesson Structure

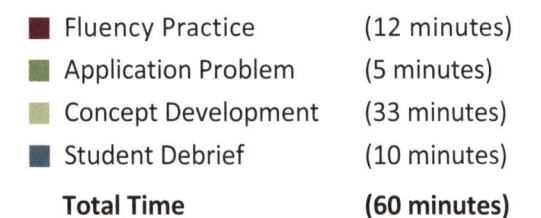

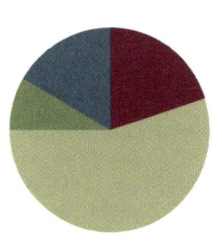

- ■ Fluency Practice   (12 minutes)
- ■ Application Problem (5 minutes)
- ■ Concept Development (33 minutes)
- ■ Student Debrief   (10 minutes)
- **Total Time**   **(60 minutes)**

Fluency Practice (12 minutes)

- Group Counting **3.OA.1**   (3 minutes)
- Tell Time on the Clock **3.MD.1**   (3 minutes)
- Minute Counting **3.MD.1**   (6 minutes)

Group Counting (3 minutes)

Note: Group counting reviews interpreting multiplication as repeated addition. Counting by sevens, eights, and nines in this activity anticipates multiplication using those units in Module 3.

Direct students to count forward and backward, occasionally changing the direction of the count using the following suggested sequence:

- Sevens to 49, emphasizing the transition from 35 to 42
- Eights to 56, emphasizing the transition from 48 to 56
- Nines to 63, emphasizing the transition from 54 to 63

Tell Time on the Clock (3 minutes)

Materials: (T) Analog clock for demonstration (S) Personal white board

Note: This activity provides additional practice with the skill of telling time to the nearest minute, taught in Lesson 3.

- T: (Show an analog demonstration clock.) Start at 12 and count by 5 minutes on the clock. (Move finger from 12 to 1, 2, 3, 4, etc., as students count.)
- S: 5, 10, 15, 20, 25, 30, 35, 40, 45, 50, 55, 60.

| A STORY OF UNITS | Lesson 4 3•2 |

T: I'll show a time on the clock. Write the time on your board. (Show 11:23.)
S: (Write 11:23.)
T: (Show 9:17.)
S: (Write 9:17.)

Repeat process, varying the hour and minute so that students read and write a variety of times to the nearest minute.

Minute Counting (6 minutes)

Note: This activity reviews the Grade 2 standard of telling and writing time to the nearest 5 minutes. Students also practice group counting strategies for multiplication in the context of time.

Use the process outlined for this activity in Lesson 1. Direct students to count by 5 minutes to 1 hour, forward and backward, naming the quarter hour and half hour intervals as such. Repeat the process:

- 6 minutes to 1 hour, naming the half hour and 1 hour intervals as such
- 3 minutes to 30 minutes, naming the quarter hour and half hour intervals as such
- 9 minutes to quarter 'til 1 hour
- 10 minutes, using the following sequence: 10 minutes, 20 minutes, 1 half hour, 40 minutes, 50 minutes, 1 hour

Application Problem (5 minutes)

Display a clock and number line as shown.

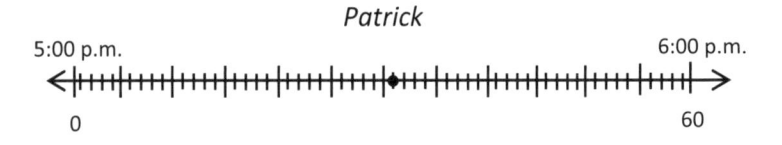

Patrick

5:00 p.m.   6:00 p.m.
←|┼┼┼┼|┼┼┼┼|┼┼┼┼|┼┼┼┼|┼┼┼┼|┼┼┼┼♦┼┼┼|┼┼┼┼|┼┼┼┼|┼┼┼┼|┼┼┼┼|→
0   60

Lilly

Patrick and Lilly start their chores at 5:00 p.m. The clock shows what time Lilly finishes. The number line shows what time Patrick finishes. Who finishes first? Explain how you know. Solve the problem without drawing a number line. You might want to visualize or use your clock template, draw a tape diagram, use words, number sentences, etc.

Patrick: 5:31
Lilly: 5:43
Patrick finishes his chores first because 5:31 comes before 5:43. I know this because I pictured Patrick's time on the clock that shows Lilly's time.

Note: This problem reviews Lesson 3, telling time to the nearest minute. This problem is used in the first example of the Concept Development to solve word problems involving minute intervals.

Lesson 4: Solve word problems involving time intervals within 1 hour by counting backward and forward using the number line and clock.

A STORY OF UNITS

Lesson 4 3•2

Concept Development (33 minutes)

Materials: (T) Analog clock for demonstration (S) Personal white board, number line (Template), clock (Lesson 3 Template)

Problem 1: Count forward and backward using a number line to solve word problems involving time intervals within 1 hour.

T: Look back at your work on today's Application Problem. We know that Lilly finished after Patrick. Let's use a number line to figure out how many more minutes than Patrick Lilly took to finish. Slip the number line Template into your personal white board.

T: Label the first tick mark 0 and the last tick mark 60. Label the hours and 5-minute intervals.

T: Plot the times 5:31 p.m. and 5:43 p.m.

T: We could count by ones from 5:31 to 5:43. Instead, discuss with a partner a more efficient way to find the difference between Patrick and Lilly's times.

S: (Discuss.)

T: Work with a partner to find the difference between Patrick's and Lilly's times.

MP.4

T: How many more minutes than Patrick did it take Lilly to finish her chores?

S: 12 minutes more.

T: What strategy did you use to solve this problem?

S: (Share possible strategies, listed below.)
- Count by ones to 5:35, by fives to 5:40, by ones to 5:43.
- Subtract 31 minutes from 43 minutes.
- Count backwards from 5:43 to 5:31.
- Know 9 minutes gets to 5:40 and 3 more minutes gets to 5:43.
- Add a ten and 2 ones.

Repeat the process with other time interval word problems, varying the unknown as suggested below.

- *Result unknown*: Start time and minutes elapsed known, end time unknown. (We started math at 10:15 a.m. We worked for 23 minutes. What time was it when we ended?)
- *Change unknown*: Start time and end time known, minutes elapsed unknown. (Leslie starts reading at 11:24 a.m. She finishes reading at 11:57 a.m. How many minutes does she read?)

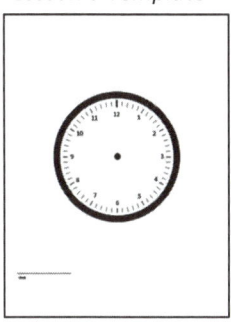

Lesson 3 Template

Number Line Template

NOTES ON MULTIPLE MEANS OF ACTION AND EXPRESSION:

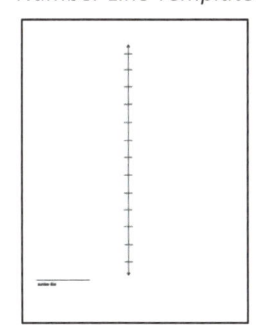

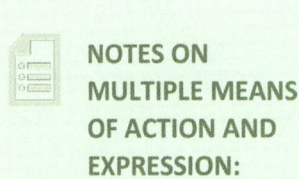

If appropriate for the class, discuss strategies for solving different problem types (*start unknown, change unknown, result unknown*). Although problem types can be solved using a range of strategies, some methods are more efficient than others depending on the unknown.

A STORY OF UNITS   Lesson 4 3•2

- *Start unknown*: End time and minutes elapsed known, start time unknown. (Joe finishes his homework at 5:48 p.m. He worked for 32 minutes. What time did he start his homework?)

Problem 2: Count forward and backward using a clock to solve word problems involving time intervals within 1 hour.

T: It took me 42 minutes to cook dinner last night. I finished cooking at 5:56 p.m. What time did I start?

T: Let's use a clock to solve this problem. Put the clock template in your board.

T: Work with your partner to draw the hands on your clock to show 5:56 p.m.

T: Talk with your partner, will you count backward or forward on the clock to solve this problem? (Allow time for discussion.)

T: Use an efficient strategy to count back 42 minutes. Write the start time on your personal white board, and as you wait for others, record your strategy.

> **NOTES ON PROBLEM TYPES:**
>
> Tables 1 and 2 in the Glossary of the *Common Core Learning Standards for Mathematics* provide a quick reference of problem types and examples.

Circulate as students work and analyze their strategies so that you can select those you would like to have shared with the whole class. Also consider the order in which strategies will be shared.

T: What time did I start making dinner?
S: 5:14 p.m.
T: I would like to ask Nina and Hakop to share their work, in that order.

Repeat the process with other time interval word problems, varying the unknown as suggested below.

- *Result unknown*: Start time and minutes elapsed known, end time unknown. (Henry starts riding his bike at 3:12 p.m. He rides for 36 minutes. What time does he stop riding his bike?)
- *Change unknown*: Start time and end time known, minutes elapsed unknown. (I start exercising at 7:12 a.m. I finish exercising at 7:53 a.m. How many minutes do I exercise?)
- *Start unknown*: End time and minutes elapsed known, start time unknown. (Cassie works on her art project for 37 minutes. She finishes working at 1:48 p.m. What time did she start working?)

> **NOTES ON MULTIPLE MEANS OF REPRESENTATION:**
>
> Students who struggle with comprehension may benefit from peers or teachers reading word problems aloud. This accommodation also provides students with the opportunity to ask clarifying questions as needed.

Problem Set (10 minutes)

Students should do their personal best to complete the Problem Set within the allotted 10 minutes. For some classes, it may be appropriate to modify the assignment by specifying which problems they work on first. Some problems do not specify a method for solving. Students should solve these problems using the RDW approach used for Application Problems.

52   Lesson 4:   Solve word problems involving time intervals within 1 hour by counting backward and forward using the number line and clock.

EUREKA MATH

Student Debrief (10 minutes)

Lesson Objective: Solve word problems involving time intervals within 1 hour by counting backward and forward using the number line and clock.

The Student Debrief is intended to invite reflection and active processing of the total lesson experience.

Invite students to review their solutions for the Problem Set. They should check work by comparing answers with a partner before going over answers as a class. Look for misconceptions or misunderstandings that can be addressed in the Debrief. Guide students in a conversation to debrief the Problem Set and process the lesson.

Any combination of the questions below may be used to lead the discussion.

- How are Problems 1 and 2 different? How did it affect the way you solved each problem?
- Did you count forward or backward to solve Problem 3? How did you decide which strategy to use?
- Discuss with a partner your strategy for solving Problem 6. What other counting strategies could you use with the clocks to get the same answer?
- Is 11:58 a.m. a reasonable answer for Problem 7? Why or why not?
- Explain to your partner how you solved Problem 8. How might you solve it without using a number line or a clock?
- How did we use counting as a strategy to problem solve today?

Exit Ticket (3 minutes)

After the Student Debrief, instruct students to complete the Exit Ticket. A review of their work will help with assessing students' understanding of the concepts that were presented in today's lesson and planning more effectively for future lessons. The questions may be read aloud to the students.

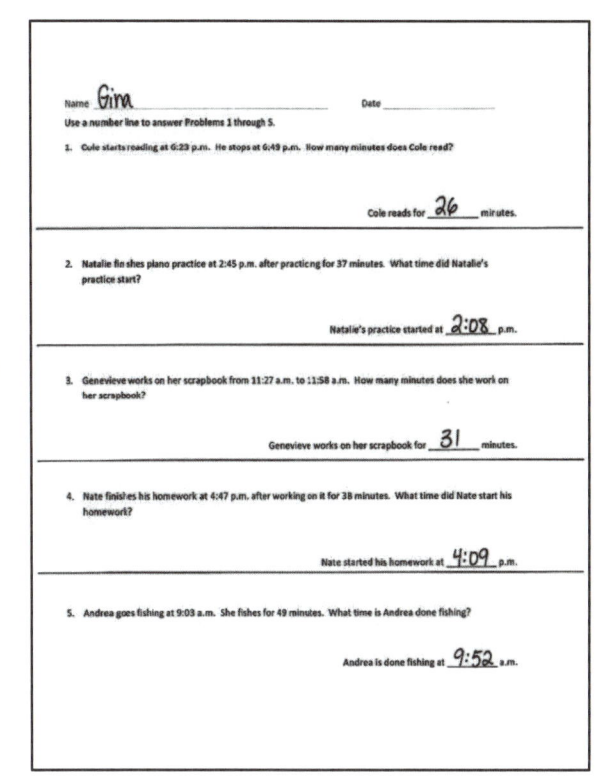

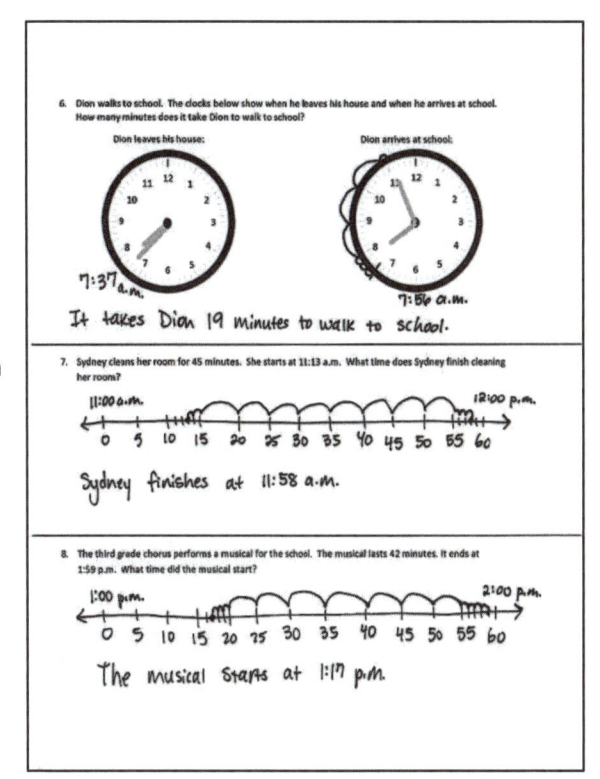

Name _____ Date _____

Use a number line to answer Problems 1 through 5.

1. Cole starts reading at 6:23 p.m. He stops at 6:49 p.m. How many minutes does Cole read?

   Cole reads for _____ minutes.

2. Natalie finishes piano practice at 2:45 p.m. after practicing for 37 minutes. What time did Natalie's practice start?

   Natalie's practice started at _____ p.m.

3. Genevieve works on her scrapbook from 11:27 a.m. to 11:58 a.m. How many minutes does she work on her scrapbook?

   Genevieve works on her scrapbook for _____ minutes.

4. Nate finishes his homework at 4:47 p.m. after working on it for 38 minutes. What time did Nate start his homework?

   Nate started his homework at _____ p.m.

5. Andrea goes fishing at 9:03 a.m. She fishes for 49 minutes. What time is Andrea done fishing?

   Andrea is done fishing at _____ a.m.

6. Dion walks to school. The clocks below show when he leaves his house and when he arrives at school. How many minutes does it take Dion to walk to school?

Dion leaves his house:

Dion arrives at school:

7. Sydney cleans her room for 45 minutes. She starts at 11:13 a.m. What time does Sydney finish cleaning her room?

8. The third-grade chorus performs a musical for the school. The musical lasts 42 minutes. It ends at 1:59 p.m. What time did the musical start?

Name _____   Date _____

Independent reading time starts at 1:34 p.m. It ends at 1:56 p.m.

1. Draw the start time on the clock below.

2. Draw the end time on the clock below.

3. How many minutes does independent reading time last?

A STORY OF UNITS

Lesson 4 Homework 3•2

Name _____ Date _____

Record your homework start time on the clock in Problem 6.

Use a number line to answer Problems 1 through 4.

1. Joy's mom begins walking at 4:12 p.m. She stops at 4:43 p.m. How many minutes does she walk?

   Joy's mom walks for _____ minutes.

2. Cassie finishes softball practice at 3:52 p.m. after practicing for 30 minutes. What time did Cassie's practice start?

   Cassie's practice started at _____ p.m.

3. Jordie builds a model from 9:14 a.m. to 9:47 a.m. How many minutes does Jordie spend building his model?

   Jordie builds for _____ minutes.

4. Cara finishes reading at 2:57 p.m. She reads for a total of 46 minutes. What time did Cara start reading?

   Cara started reading at _____ p.m.

Lesson 4: Solve word problems involving time intervals within 1 hour by counting backward and forward using the number line and clock.

A STORY OF UNITS

Lesson 4 Homework 3•2

5. Jenna and her mom take the bus to the mall. The clocks below show when they leave their house and when they arrive at the mall. How many minutes does it take them to get to the mall?

Time when they leave home:

Time when they arrive at the mall:

6. Record your homework start time:

Record the time when you finish Problems 1–5:

How many minutes did you work on Problems 1–5?

Lesson 4: Solve word problems involving time intervals within 1 hour by counting backward and forward using the number line and clock.

A STORY OF UNITS — Lesson 4 Template — 3•2

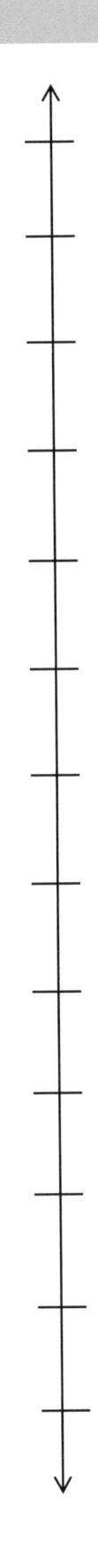

number line

Lesson 4: Solve word problems involving time intervals within 1 hour by counting backward and forward using the number line and clock.

A STORY OF UNITS   Lesson 5   3•2

Lesson 5

Objective: Solve word problems involving time intervals within 1 hour by adding and subtracting on the number line.

Suggested Lesson Structure

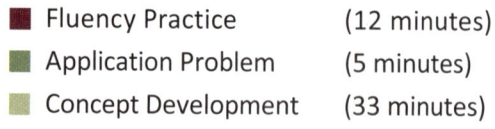

- ■ Fluency Practice   (12 minutes)
- ■ Application Problem   (5 minutes)
- ■ Concept Development   (33 minutes)
- ■ Student Debrief   (10 minutes)

   Total Time   **(60 minutes)**

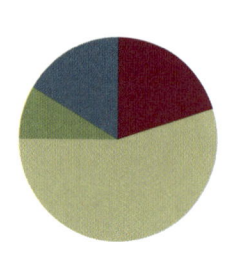

Fluency Practice (12 minutes)

- Group Counting **3.OA.1**   (3 minutes)
- Tell Time on the Clock **3.MD.1**   (3 minutes)
- Minute Counting **3.MD.1**   (6 minutes)

Group Counting (3 minutes)

Note: Group counting reviews interpreting multiplication as repeated addition. Counting by sevens, eights, and nines in this activity anticipates multiplication using those units in Module 3.

Direct students to count forward and backward, occasionally changing the direction of the count, using the following suggested sequence:

- Sevens to 56, emphasizing the transition from 49 to 56
- Eights to 64, emphasizing the transition from 56 to 64
- Nines to 72, emphasizing the transition from 63 to 72

Tell Time on the Clock (3 minutes)

Materials: (T) Analog clock for demonstration (S) Personal white board

Note: This activity provides additional practice with the newly learned skill of telling time to the nearest minute.

- T: (Show an analog demonstration clock.) Start at 12 and count by 5 minutes on the clock. (Move finger from 12 to 1, 2, 3, 4, etc., as students count.)
- S: 5, 10, 15, 20, 25, 30, 35, 40, 45, 50, 55, 60.
- T: I'll show a time on the clock. Write the time on your personal white board. (Show 5:07.)
- S: (Write 5:07.)

A STORY OF UNITS   **Lesson 5 3•2**

T: (Show 12:54.)
S: (Write 12:54.)

Repeat process, varying the hour and minute so that students read and write a variety of times to the nearest minute.

Minute Counting (6 minutes)

Note: This activity reviews the Grade 2 standard of telling and writing time to the nearest 5 minutes. Students practice group counting strategies for multiplication in the context of time.

Use the process outlined in Lesson 1. Direct students to count by 5 minutes to 1 hour, forward and backward, naming the quarter hour and half hour intervals as such. Repeat the process for the following suggested sequences:

- 3 minutes to 30 minutes, naming the quarter hour and half hour intervals as such
- 6 minutes to 1 hour, naming the half hour and 1 hour intervals as such
- 9 minutes to 45 minutes, naming the quarter hour and half hour intervals as such
  (45 minutes is named *quarter 'til 1 hour*)
- 10 minutes, using the following sequence: 10 minutes, 20 minutes, half hour, 40 minutes, 50 minutes, 1 hour

Application Problem (5 minutes)

Carlos gets to class at 9:08 a.m. He has to write down homework assignments and complete morning work before math begins at 9:30 a.m. How many minutes does Carlos have to complete his tasks before math begins?

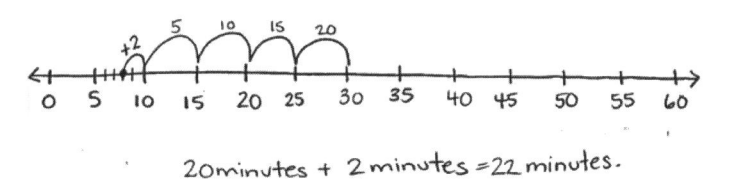

[Number line showing jumps from 8 to 10 (+2), then 5, 10, 15, 20 to 30]

20 minutes + 2 minutes = 22 minutes.
Carlos has 22 minutes to complete his tasks.

Note: This problem reviews Lesson 4 and provides a context for the problems in the Concept Development.

Encourage students to discuss how they might solve the problem using mental math strategies (e.g., count 9:18, 9:28 + 2 minutes, 2 + 20, 30 − 8).

Lesson 5: Solve word problems involving time intervals within 1 hour by adding and subtracting on the number line.

A STORY OF UNITS — Lesson 5 3•2

Concept Development (33 minutes)

Materials: (S) Personal white board, number line (Lesson 4 Template)

Lesson 4 Template

Part 1: Count forward and backward to add and subtract on the number line.

- T: Use your number line template to label the points when Carlos arrives and when math starts.
- S: (Label.)
- T: Writing down homework assignments is the first thing Carlos does when he gets to class. It takes 4 minutes. Work with your partner to plot the point that shows when Carlos finishes this first task.
- T: At what time did you plot the point?
- S: 9:12 a.m.
- T: What does the interval between 9:12 and 9:30 represent?
- S: The number of minutes it takes Carlos to finish his morning work.
- T: How can we find the number of minutes it takes Carlos to complete morning work?
- S: Count on the number line. → Count forward from 9:12 to 9:30.
- T: What addition sentence represents this problem?
- S: 12 minutes + ____ = 30 minutes.
- T: With your partner, find the number of minutes it takes Carlos to complete morning work.
- T: How many minutes did it take Carlos to finish morning work?
- S: 18 minutes.
- T: Talk with your partner. How could we have modeled that problem by counting backward?
- S: We could have started at 9:30 and counted back until we got to 9:12.
- T: What subtraction sentence represents this problem?
- S: 30 minutes − 12 minutes = 18 minutes.

Repeat the process using the following suggestions:

- Lunch starts at 12:05 p.m. and finishes at 12:40 p.m. How long is lunch?
- Joyce spends 24 minutes finding everything she needs at the grocery store. It takes her 7 minutes to pay. How long does it take Joyce to find her groceries and pay?

Part 2: Solve word problems involving time intervals within 1 hour.

- T: Gia, Carlos's classmate, gets to class at 9:11. It takes her 19 minutes to write homework assignments and complete morning work. How can we figure out if Gia will be ready to start math at 9:30?
- S: We have to find out what time Gia finishes.
- T: What do we know?
- S: We know what time Gia starts and how long it takes her to complete her tasks.
- T: What is unknown?
- S: The time that Gia finishes.

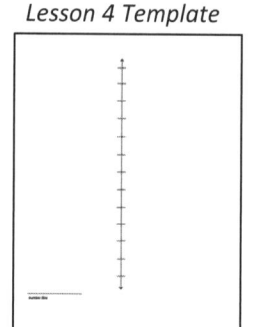

T: How can we find what time Gia finishes morning work?

S: We can start at 9:11 and add 19 minutes. → We can add 11 minutes and 19 minutes to find out how many minutes after 9:00 she finishes.

T: (Draw the model below.) Talk with your partner about why this number line shows 11 minutes + 19 minutes. (Students discuss.)

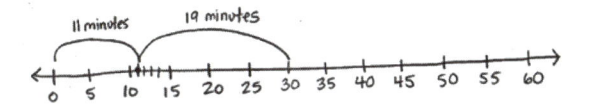

NOTES ON MULTIPLE MEANS OF ENGAGEMENT:

Relate addition on the number line with part–whole thinking. Use this connection with prior knowledge to encourage students to move from counting forward and backward toward more efficient number line representations like those modeled. Allow less confident students to verify these strategies by counting forward and backward.

T: When we add our 2 parts, 11 minutes + 19 minutes, what is our whole?

S: 30 minutes!

T: Does Gia finish on time?

S: Yes, just barely!

T: Think back to the Application Problem where Carlos gets to class at 9:08 a.m. If he left for school at 9:00 a.m., then what do the 8 minutes from 9:00 to 9:08 represent?

S: That's how long it takes Carlos to get to school.

T: We know the whole, 30 minutes, and 1 part. What does the unknown part represent?

S: The amount of time he takes to write homework and complete morning work.

T: Work with your partner to draw a number line and label the known and unknown intervals.

S: (Draw. One possible number line shown to the right.)

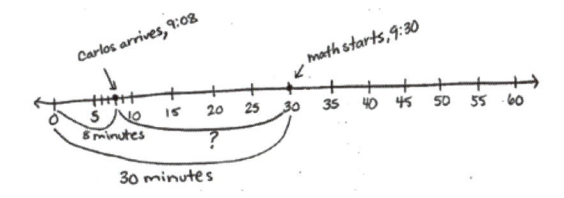

T: What is 30 minutes – 8 minutes?

S: 22 minutes!

Repeat the process using the following suggestions:

- Joey gets home at 3:25 p.m. It takes him 7 minutes to unpack and 18 minutes to have a snack before starting his homework. What is the earliest time Joey can start his homework?
- Shane's family wants to start eating dinner at 5:45 p.m. It takes him 15 minutes to set the table and 7 minutes to help put the food out. If Shane starts setting the table at 5:25 p.m., will his chores be finished by 5:45 p.m.?
- Tim gets on the bus at 8:32 a.m. and gets to school at 8:55 a.m. How long is Tim's bus ride?
- Joanne takes the same bus as Tim, but her bus ride is 25 minutes. What time does Joanne get on the bus?

NOTES ON MULTIPLE MEANS OF ENGAGEMENT:

Students who need an additional challenge can write their own word problems using real-life experiences. Encourage them to precisely time themselves during an activity and use the information to write a word problem.

A STORY OF UNITS

Lesson 5 3•2

- Davis has 3 problems for math homework. He starts at 4:08 p.m. The first problem takes him 5 minutes, and the second takes him 6 minutes. If Davis finishes at 4:23 p.m., how long does it take him to solve the last problem?

Problem Set (10 minutes)

Students should do their personal best to complete the Problem Set within the allotted 10 minutes. Depending on your class, it may be appropriate to modify the assignment by specifying which problems they work on first. Some problems do not specify a method for solving. Students should solve these problems using the RDW approach used for Application Problems.

Student Debrief (10 minutes)

Lesson Objective: Solve word problems involving time intervals within 1 hour by adding and subtracting on the number line.

The Student Debrief is intended to invite reflection and active processing of the total lesson experience.

Invite students to review their solutions for the Problem Set. They should check work by comparing answers with a partner before going over answers as a class. Look for misconceptions or misunderstandings that can be addressed in the Debrief. Guide students in a conversation to debrief the Problem Set and process the lesson.

Any combination of the questions below may be used to lead the discussion.

- Describe the process of drawing the number line for Problem 2. Explain how you labeled it. (Call on students who used different ways of thinking about and labeling parts and wholes to share.)
- How did your answer to Problem 4(a) help you solve Problem 4(b)?
- In Problem 5, you had to find a start time. How is your approach to finding a start time different from your approach to finding an end time?
- Besides a number line, what other models could you use to solve Problems 2, 4, and 5?

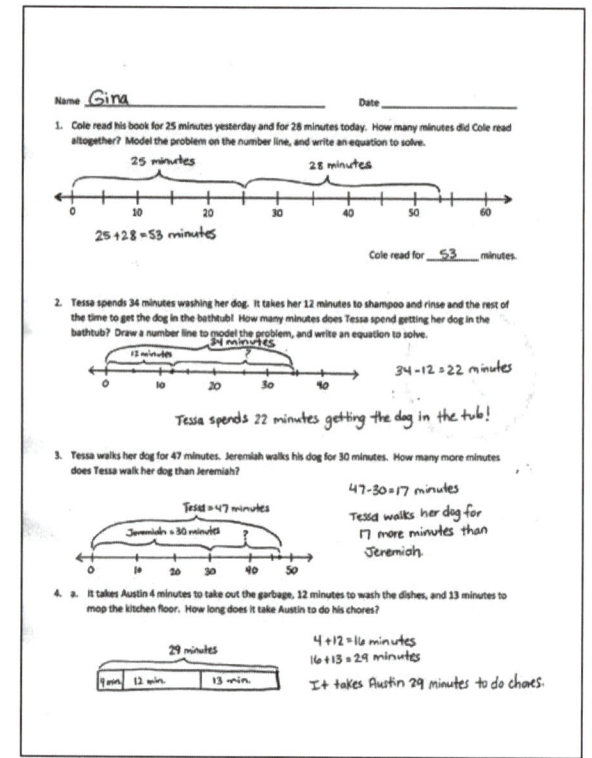

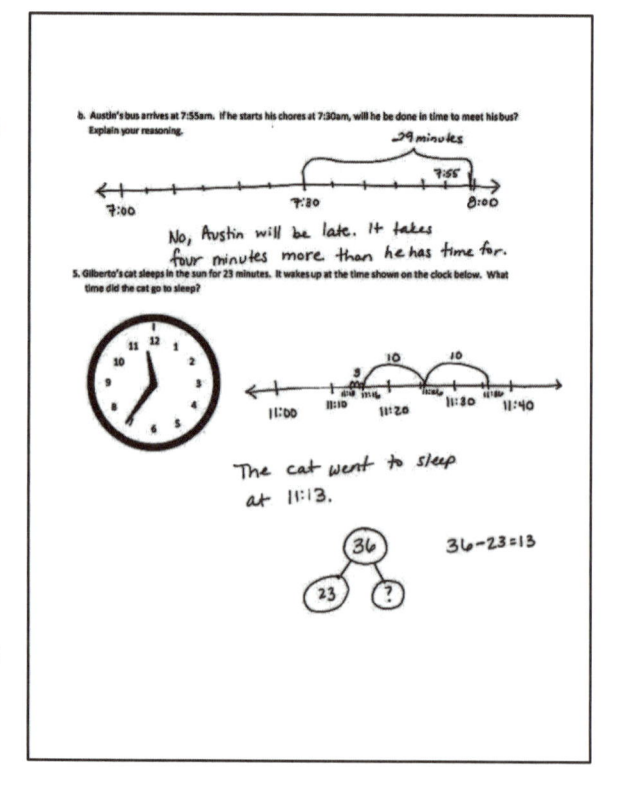

Lesson 5: Solve word problems involving time intervals within 1 hour by adding and subtracting on the number line.

Lesson 5   3•2

Exit Ticket (3 minutes)

After the Student Debrief, instruct students to complete the Exit Ticket. A review of their work will help with assessing students' understanding of the concepts that were presented in today's lesson and planning more effectively for future lessons. The questions may be read aloud to the students.

Name _____ Date _____

1. Cole read his book for 25 minutes yesterday and for 28 minutes today. How many minutes did Cole read altogether? Model the problem on the number line, and write an equation to solve.

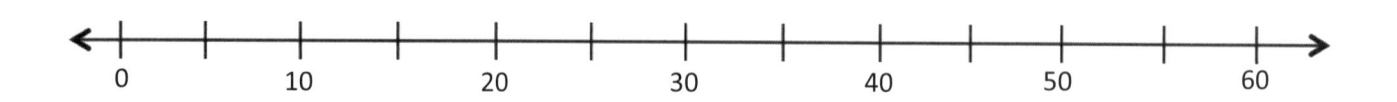

⟵|—|—|—|—|—|—|—|—|—|—|—|—|⟶
0   10   20   30   40   50   60

Cole read for _____ minutes.

2. Tessa spends 34 minutes washing her dog. It takes her 12 minutes to shampoo and rinse and the rest of the time to get the dog in the bathtub! How many minutes does Tessa spend getting her dog in the bathtub? Draw a number line to model the problem, and write an equation to solve.

3. Tessa walks her dog for 47 minutes. Jeremiah walks his dog for 30 minutes. How many more minutes does Tessa walk her dog than Jeremiah?

4. a. It takes Austin 4 minutes to take out the garbage, 12 minutes to wash the dishes, and 13 minutes to mop the kitchen floor. How long does it take Austin to do his chores?

   b. Austin's bus arrives at 7:55 a.m. If he starts his chores at 7:30 a.m., will he be done in time to meet his bus? Explain your reasoning.

5. Gilberto's cat sleeps in the sun for 23 minutes. It wakes up at the time shown on the clock below. What time did the cat go to sleep?

Name _____ Date _____

Michael spends 19 minutes on his math homework and 17 minutes on his science homework. How many minutes does Michael spend doing his homework?

Model the problem on the number line, and write an equation to solve.

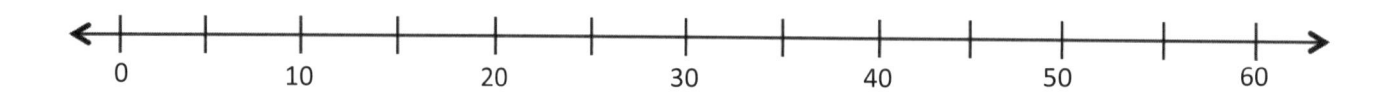

Michael spends _____ minutes on his homework.

Name _____   Date _____

1. Abby spent 22 minutes working on her science project yesterday and 34 minutes working on it today. How many minutes did Abby spend working on her science project altogether? Model the problem on the number line, and write an equation to solve.

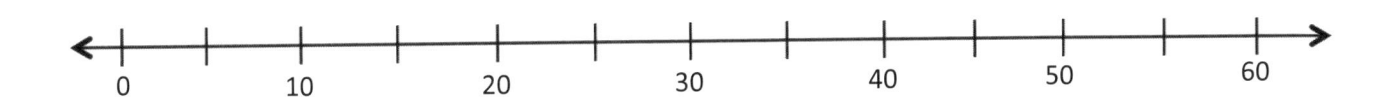

```
<---|---|---|---|---|---|---|---|---|---|---|---|--->
    0  10  20  30  40  50  60
```

Abby spent _____ minutes working on her science project.

2. Susanna spends a total of 47 minutes working on her project. How many more minutes than Susanna does Abby spend working? Draw a number line to model the problem, and write an equation to solve.

3. Peter practices violin for a total of 55 minutes over the weekend. He practices 25 minutes on Saturday. How many minutes does he practice on Sunday?

Lesson 5: Solve word problems involving time intervals within 1 hour by adding and subtracting on the number line.

4. a. Marcus gardens. He pulls weeds for 18 minutes, waters for 13 minutes, and plants for 16 minutes. How many total minutes does he spend gardening?

   b. Marcus wants to watch a movie that starts at 2:55 p.m. It takes 10 minutes to drive to the theater. If Marcus starts the yard work at 2:00 p.m., can he make it on time for the movie? Explain your reasoning.

5. Arelli takes a short nap after school. As she falls asleep, the clock reads 3:03 p.m. She wakes up at the time shown below. How long is Arelli's nap?

A STORY OF UNITS

Mathematics Curriculum

GRADE 3 • MODULE 2

Topic B
Measuring Weight and Liquid Volume in Metric Units

3.NBT.2, 3.MD.2

Focus Standards:	3.NBT.2	Fluently add and subtract within 1000 using strategies and algorithms based on place value, properties of operations, and/or the relationship between addition and subtraction.
	3.MD.2	Measure and estimate liquid volumes and masses of objects using standard units of grams (g), kilograms (kg), and liters (l). Add, subtract, multiply, or divide to solve one-step word problems involving masses or volumes that are given in the same units, e.g., by using drawings (such as a beaker with a measurement scale) to represent the problem.
Instructional Days:	6	
Coherence -Links from:	G2–M2	Addition and Subtraction of Length Units
	G2–M3	Place Value, Counting, and Comparison of Numbers to 1000
	G3–M1	Properties of Multiplication and Division and Solving Problems with Units of 2–5 and 10
-Links to:	G4–M2	Unit Conversions and Problem Solving with Metric Measurement

Lessons 6 and 7 introduce students to metric weight measured in kilograms and grams. Students learn to use digital scales as they explore these weights. They begin by holding a kilogram weight to get a sense of its weight. Then, groups of students work with scales to add rice to clear plastic zippered bags until the bags reach a weight of 1 kilogram. Once the bags reach that weight, students decompose a kilogram using ten-frames. They understand the quantity within 1 square of the ten-frame as an estimation of 100 grams. Upon that square they overlay another ten-frame, *zooming in* to estimate 10 grams. Overlaying once more leads to 1 gram. Students relate the decomposition of a kilogram to place value and the base ten system.

Throughout this two-day exploration, students reason about the size and weight of kilograms and grams in relation to one another without moving into the abstract world of conversion. They perceive the relationship between kilograms and grams as analogous to 1 meter decomposed into 100 centimeters. They build on Grade 2 estimation skills with centimeters and meters (**2.MD.3**) using metric weight. Students use scales to measure a variety of objects and learn to estimate new weights using knowledge of previously measured items. Their work with estimation in Topic B lays a foundation for rounding to estimate in the second half of the module.

In Lesson 8, students use scales to measure the weight of objects precisely, and then use those measurements to solve one-step word problems with like units. Word problems require students to add, subtract, multiply, and divide. Students apply estimation skills from Lesson 7 to reason about their solutions.

Notice that these lessons refer to *metric weight* rather than *mass*. This choice was made based on the K–5 Geometric Measurement progressions document that accompanies the CCSSM, which suggests that elementary school students may treat mass units as weight units.[1] Technically these are not equivalent, but the units can be used side by side as long as the object being measured stays on earth.[2] If students have already been introduced to the distinction between weight and mass, it may be appropriate to use the word *mass* rather than *weight*.

In Lessons 9 and 10, students measure liquid volume in liters using beakers and the vertical number line. This experience lends itself to previewing the concept and language of rounding: Students might estimate, for example, a given quantity as *halfway* between 1 and 2 or *nearer* to 2. Students use small containers to decompose 1 liter and reason about its size. This lays a conceptual foundation for Grade 4 work with milliliters and the multiplicative relationship of metric measurement units (**4.MD.1**). In these lessons, students solve one-step word problems with like units using all four operations.

Topic B culminates in solving one-step word problems with like units. Lesson 11 presents students with mixed practice, requiring students to add, subtract, multiply, and divide to find solutions to problems involving grams, kilograms, liters, and milliliters.

---

[1] Page 2 of the K–5, Geometric Measurement progression document reads, "The Standards do not differentiate between weight and mass."

[2] Page 2 of the K–5, Geometric Measurement progression document reads, "…mass is the amount of matter in an object. Weight is the force exerted on the body by gravity. On the earth's surface, the distinction is not important (on the moon, an object would have the same mass, would weight [sic] less due to the lower gravity)." To keep focused, these lessons purposefully do not introduce the distinction between weight and mass because it is not needed at this level.

| A STORY OF UNITS | Topic B | 3•2 |

A Teaching Sequence Toward Mastery of Measuring Weight and Liquid Volume in Metric Units

Objective 1: Build and decompose a kilogram to reason about the size and weight of 1 kilogram, 100 grams, 10 grams, and 1 gram.
(Lesson 6)

Objective 2: Develop estimation strategies by reasoning about the weight in kilograms of a series of familiar objects to establish mental benchmark measures.
(Lesson 7)

Objective 3: Solve one-step word problems involving metric weights within 100 and estimate to reason about solutions.
(Lesson 8)

Objective 4: Decompose a liter to reason about the size of 1 liter, 100 milliliters, 10 milliliters, and 1 milliliter.
(Lesson 9)

Objective 5: Estimate and measure liquid volume in liters and milliliters using the vertical number line.
(Lesson 10)

Objective 6: Solve mixed word problems involving all four operations with grams, kilograms, liters, and milliliters given in the same units.
(Lesson 11)

Lesson 6

Objective: Build and decompose a kilogram to reason about the size and weight of 1 kilogram, 100 grams, 10 grams, and 1 gram.

Suggested Lesson Structure

- Fluency Practice   (3 minutes)
- Concept Development   (47 minutes)
- Student Debrief   (10 minutes)
- **Total Time**   **(60 minutes)**

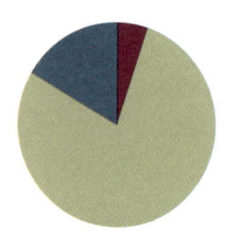

> **NOTES ON VOCABULARY IN LESSONS 6–8:**
>
> Lessons 6–8 refer to *metric weight* rather than *mass*. This choice was made based on the K–5 Geometric Measurement progressions document that accompanies the CCSSM, which suggests that elementary school students may treat mass units as weight units. Technically these are not equivalent, but the units can be used side by side as long as the object being measured stays on earth. If students have already been introduced to the distinction between weight and mass, it may be appropriate to use the word *mass* rather than *weight*. Please refer to the Topic B Opener for more information.

Fluency Practice (3 minutes)

- Tell Time on the Clock 3.MD.1   (3 minutes)

Tell Time on the Clock (3 minutes)

Materials: (T) Analog clock for demonstration
            (S) Personal white board

Note: This activity provides additional practice with the newly learned skill of telling time to the nearest minute.

T: (Show an analog demonstration clock.) Start at 12 and count by 5 minutes on the clock. (Move finger from 12 to 1, 2, 3, 4, etc., as students count.)
S: 5, 10, 15, 20, 25, 30, 35, 40, 45, 50, 55, 60.
T: I'll show a time on the clock. Write the time on your personal white board. (Show 7:13.)
S: (Write 7:13.)
T: (Show 6:47.)
S: (Write 6:47.)

Repeat process, varying the hour and minute so that students read and write a variety of times to the nearest minute.

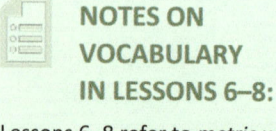

A STORY OF UNITS　　　　　　　　　　　　　　　　　　　　　　　Lesson 6　3•2

Concept Development (47 minutes)

Materials: (T) 1-kilogram weight, 1-kilogram benchmark bag of beans (S) 1-kilogram benchmark bag of beans (one per pair of students), digital metric scale, pan balance, gallon-sized sealable bag, rice, paper cup, dry-erase marker, Problem Set

Part 1: Use a pan balance to make a 1-kilogram bag of rice.

T: Today we are going to explore a **kilogram**. It's a unit used to measure weight. (Write the word *kilogram* on the board.) Whisper *kilogram* to a partner.

S: Kilogram.

T: (Pass out a 1-kilogram bag of beans to each pair of students.) You are holding 1 kilogram of beans. To record 1 kilogram, we abbreviate the word *kilogram* by writing *kg*. (Write *1 kg* on the board.) Read this weight to a partner.

S: 1 kg. → 1 kilogram.

T: (Show pan balance. See illustration in Module Overview.) This is a pan balance. Watch what happens when I put a 1-kilogram weight on one of the pans. (Turn and talk.) What will happen when I put a 1-kilogram bag of beans on the other pan?

T: (Put another bag of beans on the other side of the pan balance.) How do we know it's balanced now?

S: Both sides are the same. → Both pans have the same amount on them. That makes it balanced. → Both pans have 1 kilogram on them, so they are equal, which balances the scale.

T: (Provide pan balances, gallon-sized sealable bags, and rice.) Work with a partner.

1. Put a 1-kilogram bag of beans on one of the pans.
2. Put the empty bag on the other side, and add rice to it until the pan balance is balanced.
3. Answer Problem 1 on the Problem Set.

NOTES ON MATERIALS:

You might consider having the pre-made 1-kilogram benchmark bags hold rice, and students' benchmark bag hold beans. Beans may be easier for students to pour and clean up in case of a spill. The purpose of using these 2 different materials is for students to see that 1 kilogram is not just made using 1 particular material.

NOTES ON MULTIPLE MEANS OF REPRESENTATION:

Pre-teach new vocabulary and abbreviations whenever possible, making connections to students' prior knowledge. Highlight the similarities between *kilogram* and *kg* to aid comprehension and correct usage.

NOTES ON MULTIPLE MEANS OF ENGAGEMENT:

Provide a checklist of the steps to support students in monitoring their own progess.

Lesson 6: Build and decompose a kilogram to reason about the size and weight of 1 kilogram, 100 grams, 10 grams, and 1 gram.

Part 2: Decompose 1 kilogram.

Students work in pairs.

T: Be sure your bag is sealed, and then lay it flat on your desk. Move the rice to smooth it out until it fills the bag.

T: Using your dry-erase marker, estimate to draw a ten-frame that covers the whole bag of rice. (Ten-frame drawn on the bag on the right.)

T: The whole bag contains 1 kilogram of rice. We just decomposed the rice into 10 equal parts. These equal parts can be measured with a smaller unit of weight called **grams**. (Write *grams* on the board.) Whisper the word *grams* to your partner.

S: Grams.

T: Each part of the ten-frame is about 100 grams of rice. To record 100 grams, we can abbreviate using the letter *g*. (Write *100 g* on the board.) Write 100 g in each part of the ten-frame.

T: How many hundreds are in 1 kilogram of rice?

S: 10 hundreds!

T: Let's skip-count hundreds to find how many grams of rice are in the whole bag. Point to each part of the ten-frame as we skip-count.

S: (Point and skip-count.) 100, 200, 300, 400, 500, 600, 700, 800, 900, 1000.

T: How many grams of rice are in the whole bag?

S: 1000 grams!

T: One kilogram of rice is the same as 10 hundreds, or 1000 grams, of rice.

T: A digital scale helps us measure the weight of objects. Let's use it to measure 100 grams of rice. To measure weight on this scale, you read the number on the display screen. There is a *g* next to the display screen which means that this scale measures in grams. Put an empty cup on your digital scale. Carefully scoop rice from your bag into the cup until the scale reads 100 g.

T: How many grams are still in your bag?

S: 900 grams.

T: How many grams are in your cup?

S: 100 grams.

T: Turn and talk to a partner, will your bag of rice balance the pan balance with the 1-kilogram bag of beans? Why or why not?

T: Check your prediction by using the pan balance to see if the bag of rice balances with the bag of beans.

S: (Use pan balance to see that the bags are not balanced anymore.)

Decompose 1 kg into 10 groups of 100 g.

Decompose 100 g into 10 groups of 10 g.

Lesson 6: Build and decompose a kilogram to reason about the size and weight of 1 kilogram, 100 grams, 10 grams, and 1 gram.

T: Carefully set the cup of rice on the same pan as the bag of rice. Is it balanced now?
S: Yes, because both sides are 1 kilogram!
T: Pour the rice from the cup back into the bag. How many grams are in the bag?
S: 1000 grams.
T: Answer Problem 2 on your Problem Set.

Follow the same process to further decompose:

- Decompose 100 grams into 10 groups of 10 grams by drawing a new ten-frame within 1 part of the first ten-frame (shown to the right). Use the digital scale to scoop 100 grams into a cup again and then scoop 10 grams into another cup. How many grams are left in the first cup? How many grams are in the smaller cup? Students pour the rice back into the bag and answer Problem 3.

- Decompose 10 grams into 10 groups of 1 gram by drawing a new ten-frame within 1 part of the second ten-frame (shown to the right). Have a discussion about the difficulty of weighing 1 gram using the previous method. Students answer Problem 4.

Decompose 10 g into 10 groups of 1 g.

Problem Set (5 minutes)

Problems 1–4 in the Problem Set are intended to be completed during the Concept Development. Students can use this time to complete Problem 5.

Student Debrief (10 minutes)

Lesson Objective: Build and decompose a kilogram to reason about the size and weight of 1 kilogram, 100 grams, 10 grams, and 1 gram.

The Student Debrief is intended to invite reflection and active processing of the total lesson experience.

Invite students to review their solutions for the Problem Set. They should check work by comparing answers with a partner before going over answers as a class. Look for misconceptions or misunderstandings that can be addressed in the Debrief. Guide students in a conversation to debrief the Problem Set and process the lesson.

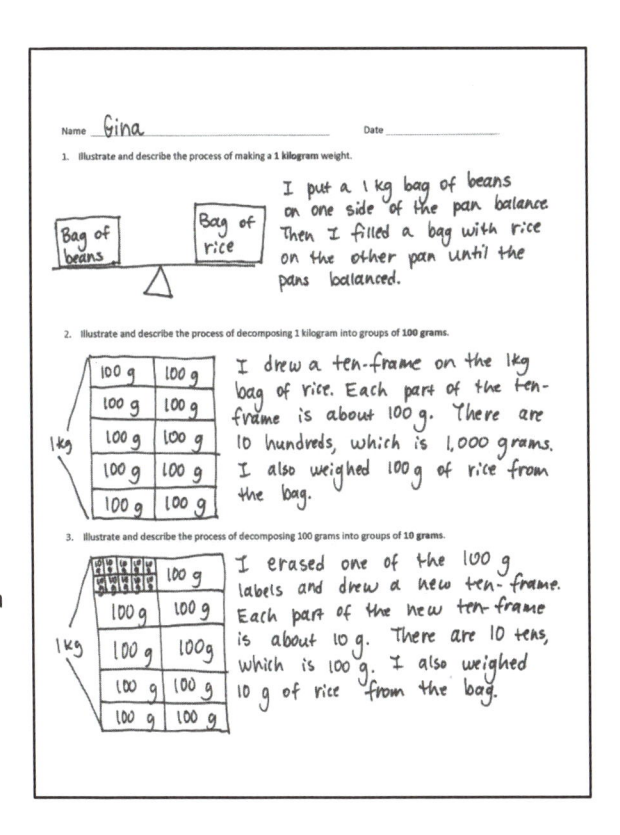

A STORY OF UNITS   Lesson 6 3•2

Any combination of the questions below may be used to lead the discussion.

- How are the units **kilogram** and **gram** similar? How are they different?
- Explain to a partner how you used a pan balance to create a bag of rice that weighed 1 kilogram.
- Could we have used the digital scale to create a bag of rice that weighs 1 kilogram? Why or why not?
- How many equal parts were there when you decomposed 1 kilogram into groups of 100 grams? 100 grams into groups of 10 grams? 10 grams into groups of 1 gram? How does this relationship help you answer Problem 5?

MP.6
- What new math vocabulary did we use today to communicate precisely about weight?
- At the beginning of our lesson, we used a number bond to show an hour in two parts that together made the whole. How did we also show parts that together made a whole kilogram?

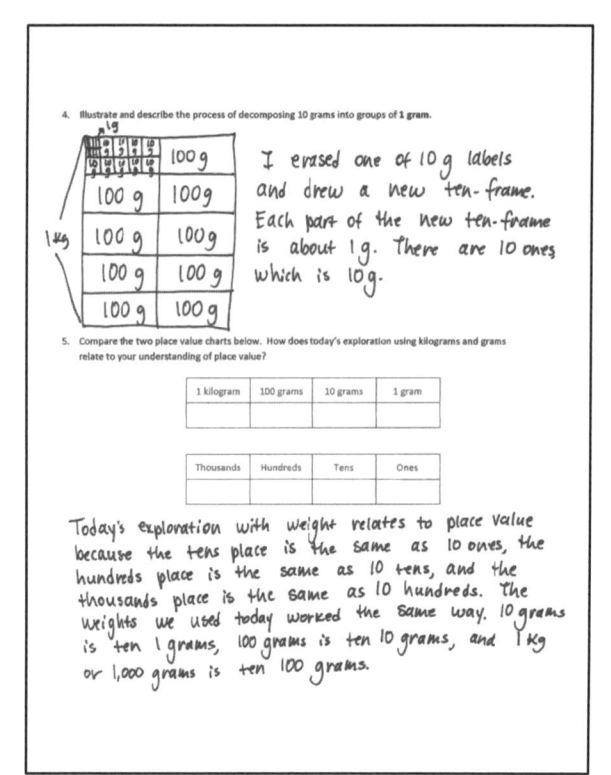

Exit Ticket (3 minutes)

After the Student Debrief, instruct students to complete the Exit Ticket. A review of their work will help with assessing students' understanding of the concepts that were presented in today's lesson and planning more effectively for future lessons. The questions may be read aloud to the students.

Lesson 6: Build and decompose a kilogram to reason about the size and weight of 1 kilogram, 100 grams, 10 grams, and 1 gram.

Name _____   Date _____

1. Illustrate and describe the process of making a 1-kilogram weight.

2. Illustrate and describe the process of decomposing 1 kilogram into groups of 100 grams.

3. Illustrate and describe the process of decomposing 100 grams into groups of 10 grams.

4. Illustrate and describe the process of decomposing 10 grams into groups of 1 gram.

5. Compare the two place value charts below. How does today's exploration using kilograms and grams relate to your understanding of place value?

1 kilogram	100 grams	10 grams	1 gram

Thousands	Hundreds	Tens	Ones

Name _____ Date _____

Ten bags of sugar weigh 1 kilogram. How many grams does each bag of sugar weigh?

Name _____ Date _____

1. Use the chart to help you answer the following questions:

1 kilogram	100 grams	10 grams	1 gram

   a. Isaiah puts a 10-gram weight on a pan balance. How many 1-gram weights does he need to balance the scale?

   b. Next, Isaiah puts a 100-gram weight on a pan balance. How many 10-gram weights does he need to balance the scale?

   c. Isaiah then puts a kilogram weight on a pan balance. How many 100-gram weights does he need to balance the scale?

   d. What pattern do you notice in Parts (a–c)?

2. Read each digital scale. Write each weight using the word *kilogram* or *gram* for each measurement.

3 kg

6 kg

450 g

_____

_____

_____

907 g

11 kg

1 kg

_____

_____

_____

Lesson 7

Objective: Develop estimation strategies by reasoning about the weight in kilograms of a series of familiar objects to establish mental benchmark measures.

Suggested Lesson Structure

- Fluency Practice   (10 minutes)
- Application Problem   (3 minutes)
- Concept Development   (37 minutes)
- Student Debrief   (10 minutes)
- **Total Time**   **(60 minutes)**

Fluency Practice (10 minutes)

- Group Counting **3.OA.1**   (4 minutes)
- Decompose 1 Kilogram **3.MD.2**   (4 minutes)
- Gram Counting **3.MD.2**   (2 minutes)

Group Counting (4 minutes)

Note: Group counting reviews interpreting multiplication as repeated addition. The counting by groups in this activity reviews foundational strategies for multiplication from Module 1 and anticipates Module 3.

Direct students to count forward and backward, occasionally changing the direction of the count using the following suggested sequence:

- Threes to 30
- Fours to 40
- Sixes to 60
- Sevens to 70, emphasizing the transition from 63 to 70
- Eights to 80, emphasizing the transition from 72 to 80
- Nines to 90, emphasizing the transition from 81 to 90

As students improve with skip-counting, e.g., 7, 14, 21, 28, etc., have them keep track of how many groups they have counted on their fingers. Keep asking them to say the number of groups, e.g., "24 is how many threes?" "63 is how many sevens?"

A STORY OF UNITS   Lesson 7   3•2

Decompose 1 Kilogram (4 minutes)

Materials: (S) Personal white board

Note: Decomposing 1 kilogram using a number bond helps students relate part–whole thinking to measurement concepts. It also sets the foundation for work with fractions.

- T: (Project a number bond with 1 kg written as the whole.) There are 1,000 grams in 1 kilogram.
- T: (Write 900 grams as one of the parts.) On your personal white board, write a number bond filling in the unknown part.
- S: (Draw number bond with 100 g, completing the unknown part.)

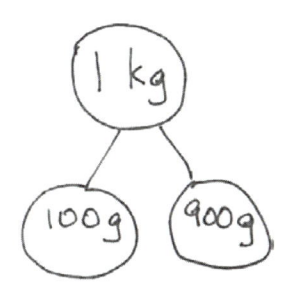

Continue with the following possible sequence: 500 g, 700 g, 400 g, 600 g, 300 g, 750 g, 650 g, 350 g, 250 g, 850 g, and 150 g. Do as many as possible within the four minutes allocated for this activity.

Gram Counting (2 minutes)

Note: This activity reviews Lesson 6 and lays a foundation for Grade 4 when students compose compound units of kilograms and grams.

- T: There are 1,000 grams in 1 kilogram. Count by 100 grams to 1 kilogram.
- S: 100 grams, 200 grams, 300 grams, 400 grams, 500 grams, 600 grams, 700 grams, 800 grams, 900 grams, 1 kilogram.

Application Problem (3 minutes)

Justin put a 1-kilogram bag of flour on one side of a pan balance. How many 100-gram bags of flour does he need to put on the other pan to balance the scale?

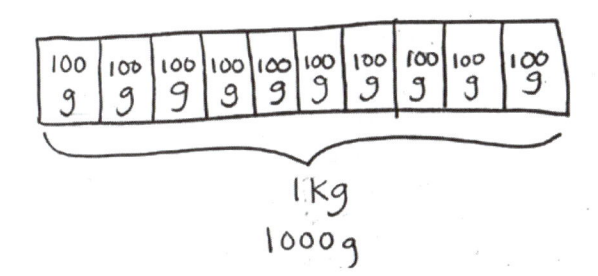

$100g \times 10 = 1000g$

Justin has to put 10 bags of flour on the other pan to balance the scale.

Note: This problem reviews the decomposition of 1 kilogram and the vocabulary words *kilogram* and *gram* from Lesson 6. Student work shown above is exemplary work. Students may also solve with repeated addition or skip-counting. Invite discussion by having students share a variety of strategies.

Lesson 7:   Develop estimation strategies by reasoning about the weight in kilograms of a series of familiar objects to establish mental benchmark measures.

A STORY OF UNITS
Lesson 7 3•2

Concept Development (37 minutes)

Materials: (T) Digital scale in grams (S) Metric spring scale

Part 1: Become familiar with scales.

Draw or project spring scales shown below on the board.

- T: (Show spring scale. See illustration in Module Overview.) This is a spring scale. There is a *g* on this scale. That means it can be used to measure grams. Other spring scales measure in kilograms. I've drawn some on the board. (See examples below.)
- T: (Point to the first drawing.) This scale shows the weight of a bowl of apples. Each interval on this scale represents 1 kilogram. How much does the bowl of apples weigh?
- S: 3 kilograms.
- T: Talk to your partner. Where would the arrow point if it weighed 1 kilogram? 4 kilograms?
- T: Look at the next scale, weighing rice. Each interval on this scale represents 500 grams. How much does the bag of rice weigh?
- S: 1,000 grams. → 1 kilogram.
- T: Talk to your partner about how this scale would show 3 kilograms. What about 5 kilograms?
- T: On the last scale, 5 intervals represent 500 grams. How much does 1 interval represent?
- S: 100 grams!
- T: Let's count grams on this scale to find 1 kilogram. (Move finger and count 100 grams, 200 grams, 300 grams, etc.)
- T: Where is 1 kilogram on this scale? 200 grams?
- S: (Discuss.)

> **NOTES ON SCALES:**
>
> The scales available to you may be different from those used in the vignette. Change the directions as necessary to match the tools at your disposal.
>
> Unlike a clock, a spring scale may be labeled in different ways. This adds the complexity that the value of the whole may change, therefore changing the value of the interval.

- T: (Pass out spring scales that measure in grams.) This scale is labeled in intervals of 200. Skip-count by two-hundreds to find how many grams the scale can measure.
- S: (Point and skip-count.) 200, 400, 600, 800, 1,000, 1,200, 1,400, 1,600, 1,800, 2,000.

Lesson 7: Develop estimation strategies by reasoning about the weight in kilograms of a series of familiar objects to establish mental benchmark measures.

A STORY OF UNITS   **Lesson 7** 3•2

T: This scale can measure 2,000 grams. That means that each tick mark represents 20 grams. Working with a partner, start at 0 and skip-count by twenties to find the 100-gram mark on this scale.

S: (Work with a partner and skip-count to 100.) 20, 40, 60, 80, 100.

Continue having students locate weights on this scale with the following possible sequence: 340 g, 880 g, and 1,360 g.

T: To accurately measure objects that weigh less than 20 grams, we are going to use a digital scale. (Show digital scale.) Remember from yesterday, to measure weight on this scale, you read the number on the display screen. (Point to display screen.) There is a *g* next to the display screen which means that this scale measures in grams. (Model measuring.)

T: We'll use both a spring scale and a digital scale in today's exploration.

Part 2: Exploration Activity

Students begin to use estimation skills as they explore the weight of 1 kilogram. In one hand, they hold a 1-kilogram weight, and with the other, they pick up objects around the room that they think weigh about the same as 1 kilogram. Students determine whether the objects weigh *less than*, *more than*, or *about the same as* 1 kilogram. Encourage students to use the italicized comparative language. Next, they weigh the objects using scales and compare their estimates with precise measurements. They repeat this process using 100-gram, 10-gram, and 1-gram weights.

Demonstrate the process of using the kilogram weight. For example, pick up the 1-kilogram weight and a small paperback book. Think out loud so students can hear you model language and thinking to estimate that the book weighs less than 1 kilogram. Repeat the process with an object that weighs *more than* and *about the same as* 1 kilogram.

> **NOTES ON MULTIPLE MEANS OF ACTION AND EXPRESSION:**
>
> Comparative language is often difficult for English language learners. Depending on your class, pre-teach the vocabulary and provide students with sentence frames.

Problem Set (20 minutes)

Materials: (S) 1 kg, 100 g, 10 g, and 1 g weights (or pre-measured and labeled bags of rice corresponding to each measurement), spring scale that measures up to 2,000 grams, metric digital scale

Side 1 of the Problem Set is used for the lesson's exploration. Students should complete Side 2 independently or with a partner.

> **NOTES ON MULTIPLE MEANS OF REPRESENTATION:**
>
> Review directions with students before they begin. Use prompts to get them to stop and think before moving to the next step, e.g., "Who remembers the next step?"

Lesson 7:   Develop estimation strategies by reasoning about the weight in kilograms of a series of familiar objects to establish mental benchmark measures.

A STORY OF UNITS

Lesson 7 3•2

Student Debrief (10 minutes)

Lesson Objective: Develop estimation strategies by reasoning about the weight in kilograms of a series of familiar objects to establish mental benchmark measures.

The Student Debrief is intended to invite reflection and active processing of the total lesson experience.

Invite students to review their solutions for the Problem Set. They should check work by comparing answers with a partner before going over answers as a class. Look for misconceptions or misunderstandings that can be addressed in the Debrief. Guide students in a conversation to debrief the Problem Set and process the lesson.

Any combination of the questions below may be used to lead the discussion.

- How did you use the 1-kilogram, 100-gram, 10-gram, and 1-gram weights to help you estimate the weights of objects in the classroom?
- Today you used a spring scale and a digital scale to measure objects. How are these scales used differently than the pan balance from yesterday's lesson?
- Did anyone find an object that weighs exactly 1 kilogram? What object? (Repeat for 100 grams, 10 grams, and 1 gram.)
- Look at Problem D. List some of the actual weights you recorded (there should be a huge variation in weights for this problem). Why do you suppose there are a small number of weights very close to 1 gram?
- Discuss Problem E with a partner. How did you determine which estimation was correct for each object?
- Discuss Problem F. (This problem anticipates the introduction of liters in Lessons 9 and 10, hinting at the weight equivalence of 1 liter of water and 1 kilogram.)
- Problem G reminds me of a riddle I know: What weighs more, 1 kilogram of bricks or 1 kilogram of feathers? Think about the relationship between the beans and rice in Problem G to help you answer this riddle.

Exit Ticket (3 minutes)

After the Student Debrief, instruct students to complete the Exit Ticket. A review of their work will help with assessing students' understanding of the concepts that were presented in today's lesson and planning more effectively for future lessons. The questions may be read aloud to the students.

Name _____ Date _____

Work with a partner. Use the corresponding weights to estimate the weight of objects in the classroom. Then, check your estimate by weighing on a scale.

A.

Objects that Weigh About **1 Kilogram**	Actual Weight

B.

Objects that Weigh About **100 Grams**	Actual Weight

C.

Objects that Weigh About **10 Grams**	Actual Weight

D.

Objects that Weigh About **1 Gram**	Actual Weight

E. Circle the correct unit of weight for each estimation.

1. A box of cereal weighs about 350 (grams / kilograms).

2. A watermelon weighs about 3 (grams / kilograms).

3. A postcard weighs about 6 (grams / kilograms).

4. A cat weighs about 4 (grams / kilograms).

5. A bicycle weighs about 15 (grams / kilograms).

6. A lemon weighs about 58 (grams / kilograms).

F. During the exploration, Derrick finds that his bottle of water weighs the same as a 1-kilogram bag of rice. He then exclaims, "Our class laptop weighs the same as 2 bottles of water!" How much does the laptop weigh in kilograms? Explain your reasoning.

G. Nessa tells her brother that 1 kilogram of rice weighs the same as 10 bags containing 100 grams of beans each. Do you agree with her? Explain why or why not.

Name _____ Date _____

1. Read and write the weights below. Write the word *kilogram* or *gram* with the measurement.

   _____   _____

2. Circle the correct unit of weight for each estimation.

   a. An orange weighs about 200 (grams / kilograms).

   b. A basketball weighs about 624 (grams / kilograms).

   c. A brick weighs about 2 (grams / kilograms).

   d. A small packet of sugar weighs about 4 (grams / kilograms).

   e. A tiger weighs about 190 (grams / kilograms).

Name _____   Date _____

1. Match each object with its approximate weight.

   [water bottle] •   • 100 grams

   [paperclip] •   • 10 grams

   [4 pennies] •   • 1 gram

   [apple] •   • 1 kilogram

2. Alicia and Jeremy weigh a cell phone on a digital scale. They write down 113 but forget to record the unit. Which unit of measurement is correct, grams or kilograms? How do you know?

3. Read and write the weights below. Write the word *kilogram* or *gram* with the measurement.

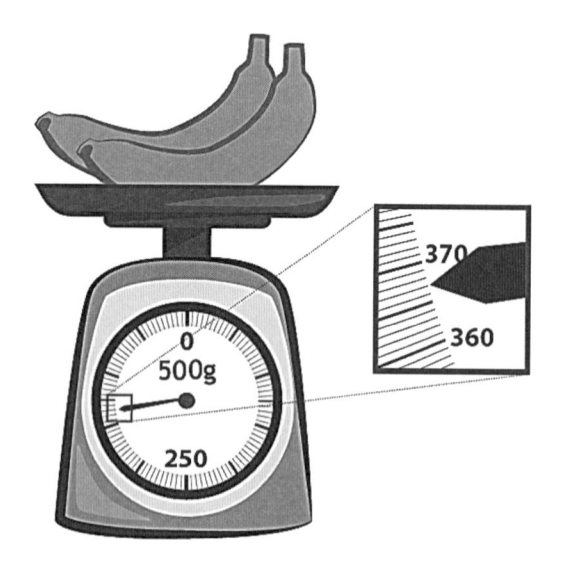

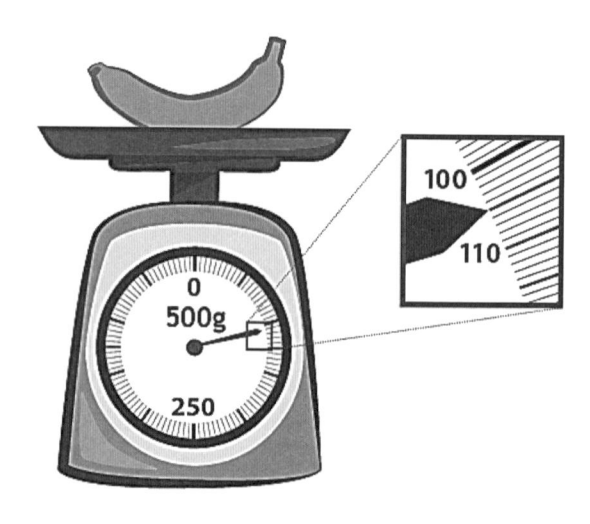

Lesson 8

Objective: Solve one-step word problems involving metric weights within 100 and estimate to reason about solutions.

Suggested Lesson Structure

- ■ Fluency Practice   (8 minutes)
- ■ Concept Development   (42 minutes)
- ■ Student Debrief   (10 minutes)
- **Total Time**   **(60 minutes)**

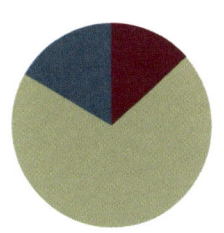

Fluency Practice (8 minutes)

- Divide Grams and Kilograms **3.MD.2**   (2 minutes)
- Determine the Unit of Measure **3.MD.2**   (2 minutes)
- Group Counting **3.OA.1**   (4 minutes)

Divide Grams and Kilograms (2 minutes)

Note: This activity reviews the decomposition of 1 kg, 100 g, and 10 g from Lesson 6, as well as division skills using units of 10 from Module 1.

   T: (Project 10 g ÷ 10 = ___.) Read the division sentence.
   S: 10 grams ÷ 10 = 1 gram.

Continue with the following possible sequence: 100 g ÷ 10 and 1,000 g ÷ 10.

Determine the Unit of Measure (2 minutes)

Note: This activity reviews the difference in size between grams and kilograms as units of measurement from Lesson 7.

   T: I'll name an object. You say if it should be measured in grams or kilograms. Apple.
   S: Grams.

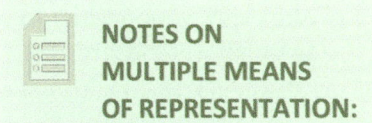

Continue with the following possible sequence: carrot, dog, pencil, classroom chair, car tire, and paper clip.

> **NOTES ON MULTIPLE MEANS OF REPRESENTATION:**
>
> Provide a visual. Use a place value chart to model the division, making an explicit connection between metric weight measurement and the base ten system.

A STORY OF UNITS   Lesson 8   3•2

Group Counting (4 minutes)

Note: Group counting reviews interpreting multiplication as repeated addition. The group counting in this activity reviews foundational multiplication strategies from Module 1 and anticipates units used in Module 3.

Direct students to count forward and backward, occasionally changing the direction of the count:

- Threes to 30
- Fours to 40
- Sixes to 60
- Sevens to 70
- Eights to 80
- Nines to 90

As students become more fluent with skip-counting by a particular unit, have them track the number of groups counted on their fingers.

Concept Development (42 minutes)

Materials: (T) Spring scale, digital scale (S) Spring scales that measure grams, personal white board, 1-kg bag of rice, beans (baggie weighing 28 g per pair), popcorn kernels (baggie weighing 36 g per pair)

Problem 1: Solve one-step word problems using addition.

Pairs of students have spring scales and baggies of beans and popcorn kernels.

- T: Let's use spring scales to weigh our beans and kernels. Should we use grams or kilograms?
- S: Grams!
- T: Compare the feel of the beans and the popcorn kernels. Which do you think weighs more?
- S: (Pick up bags and estimate.)
- T: Work with your partner to weigh the beans and kernels. Record the measurements on your personal white board.
- S: (Weigh and record. Beans weigh about 28 grams, and kernels weigh about 36 grams.)
- T: Was your estimation correct? Tell your partner.
- S: (Share.)
- T: Let's add to find the total weight of the beans and kernels. Solve the problem on your personal white board.
- S: (Solve.)
- T: I noticed someone used a simplifying strategy to add. She noticed that 28 grams is very close to 30 grams. Thirty is an easier number to add than 28. Watch how she made a ten to add. (Model sequence on the next page.)

Lesson 8: Solve one-step word problems involving metric weights within 100 and estimate to reason about solutions.

A STORY OF UNITS   Lesson 8 3•2

$$28\,g + 36\,g =$$
$$\quad\quad\quad\;\;/\backslash$$
$$\quad\quad\quad 2\;\;34$$

$$30\,g + 34\,g = 64\,g$$
$$\quad\quad\quad\;\;/\backslash$$
$$\quad\quad\quad 30\;\;4$$

NOTES ON MULTIPLE MEANS OF ACTION AND EXPRESSION:

Depending on timing and the variety of strategies used to solve, consider selecting a few students to share their work.

T: How might this strategy help us solve other similar problems using mental math?

S: From 28 it was easy to make 30, so I guess when there's a number close to a ten, like 39 or 58, we can just get 1 or 2 out of the other number to make a ten. → Yeah, it is easy to add tens like 20, 30, 40. → So, 49 + 34 becomes 49 + 1 + 33, then 50 + 33 = 83. → Oh! One just moved from 34 to 49!

T: Tell your partner how we could have used our scales to find the total weight.

S: We could have weighed the beans and kernels together!

T: Do that now to check your calculation.

Problem 2: Solve one-step word problems using subtraction.

T: (Project this *compare lesser (smaller) unknown* problem with *result unknown* problem.) Lindsey wants to ride the roller coaster. The minimum weight to ride is 32 kilograms. She weighs 14 kilograms less than the required weight. How many kilograms does Lindsey weigh?

T: Work with your partner to draw and write an equation to model the problem.

S: (Model.)

T: How will you solve? Why will you do it that way?

S: (Discuss. Most will likely agree on subtraction: 32 kg – 14 kg.)

T: Talk with your partner about how you might use tens to make a simplifying strategy for solving.

S: How about 32 – 10 – 4? → Or we could break 14 into 10 + 2 + 2. Then it's easy to do 32 – 2 – 10 – 2.

T: Solve the problem now. (Select one to two pairs of students to demonstrate their work.)

As time allows, repeat the process.

NOTES ON MULTIPLE MEANS OF ENGAGEMENT:

Adjust students' level of independence to provide an appropriate challenge. It may work well for some students to work on their own and for others to work in pairs.

An intermediate level of scaffolding might be for students to work in pairs, with each partner solving one of the two problems. Then partners can check each other's work and compare strategies. *Result unknown* problems are usually easiest. Assign problems strategically.

- *Take from with result unknown*: Ms. Casallas buys a new cabinet for the classroom. It comes in a box that weighs 42 kilograms. Ms. Casallas unpacks pieces that total 16 kilograms. How much does the box weigh now?
- *Take from with change unknown*: Mr. Flores weighs 73 kilograms. After exercising every day for 6 weeks, he loses weight. Now he weighs 67 kilograms. How much weight did he lose?

Lesson 8: Solve one-step word problems involving metric weights within 100 and estimate to reason about solutions.

Problem 3: Solve one-step word problems using multiplication.

T: Let's use a digital scale to measure the weight of Table 1's supply box. (Model weighing.)

T: It weighs about 2 kilograms. Talk with your partner. Is it reasonable to suppose that the supply boxes at each table weigh about 2 kilograms?

S: No, because ours has more crayons than Table 1's. → But it's not very many crayons, and they don't weigh very much. Besides, the teacher said *about* 2 kilograms. → It's reasonable because they are the same box, and they all have almost the exact same things in them.

MP.4

T: How are we using a simplifying strategy by supposing that each of the boxes weighs about 2 kilograms?

S: It's simpler because we don't have to weigh everything. → It's simplifying because then we can just multiply the number of boxes times 2 kilograms. Multiplying by two is easier than adding a bunch of different numbers together.

T: Partner A, model and solve this problem. Explain your solution to Partner B. Partner B, check your friend's work. Then write and solve a different multiplication sentence to show the problem. Explain to, or model for, Partner A why your multiplication sentence makes sense, too.

S: (Partner A models and writes 6 × 2. Partner B checks work and writes and explains 2 × 6.)

As time allows, repeat the process with the following suggested *equal measures with unknown product* problems.

- Jerry buys 3 bags of groceries. Each bag weighs 4 kilograms. How many kilograms do Jerry's grocery bags weigh in all?
- A dictionary weighs 3 kilograms. How many kilograms do 9 dictionaries weigh?

Problem 4: Solve one-step word problems using division.

T: (Project this *equal measures with group size unknown* problem.) Eight chairs weigh 24 kilograms. What is the weight of 1 chair? Work with your partner to model or write an equation to represent the problem.

S: (Model and/or write 24 ÷ 8 = ____.)

T: What will be your strategy for solving?

S: We can skip-count by eights just like we practiced in today's fluency activity!

As time allows, repeat the process.

- *Equal measures with group size unknown*:
  Thirty-six kilograms of apples are equally distributed into 4 crates. What is the weight of each crate?
- *Equal measures with number of groups unknown*: A tricycle weighs 8 kilograms. The delivery truck is almost full but can hold 40 kilograms more. How many more tricycles can the truck hold?

Problem Set (10 minutes)

Students should do their personal best to complete the Problem Set within the allotted 10 minutes. For some classes, it may be appropriate to modify the assignment by specifying which problems they work on first. Some problems do not specify a method for solving. Students should solve these problems using the RDW approach used for Application Problems.

A STORY OF UNITS   Lesson 8   3•2

Student Debrief (10 minutes)

Lesson Objective: Solve one-step word problems involving metric weights within 100 and estimate to reason about solutions.

The Student Debrief is intended to invite reflection and active processing of the total lesson experience. Invite students to review their solutions for the Problem Set.

They should check work by comparing answers with a partner before going over answers as a class. Look for misconceptions or misunderstandings that can be addressed in the Debrief. Guide students in a conversation to debrief the Problem Set and process the lesson.

Any combination of the questions below may be used to lead the discussion.

- How did your tape diagrams change in Problems 2(a) and 2(b)?
- Explain to your partner the relationship between Problem 2(a) and Problem 2(b).
- How did today's Fluency Practice help you with problem solving during the Concept Development?
- Select students to share simplifying strategies or mental math strategies they used to solve problems in the Problem Set. If no one used a special strategy or mental math, brainstorm about alternative ways for solving Problem 2.

Exit Ticket (3 minutes)

After the Student Debrief, instruct students to complete the Exit Ticket. A review of their work will help with assessing students' understanding of the concepts that were presented in today's lesson and planning more effectively for future lessons. The questions may be read aloud to the students.

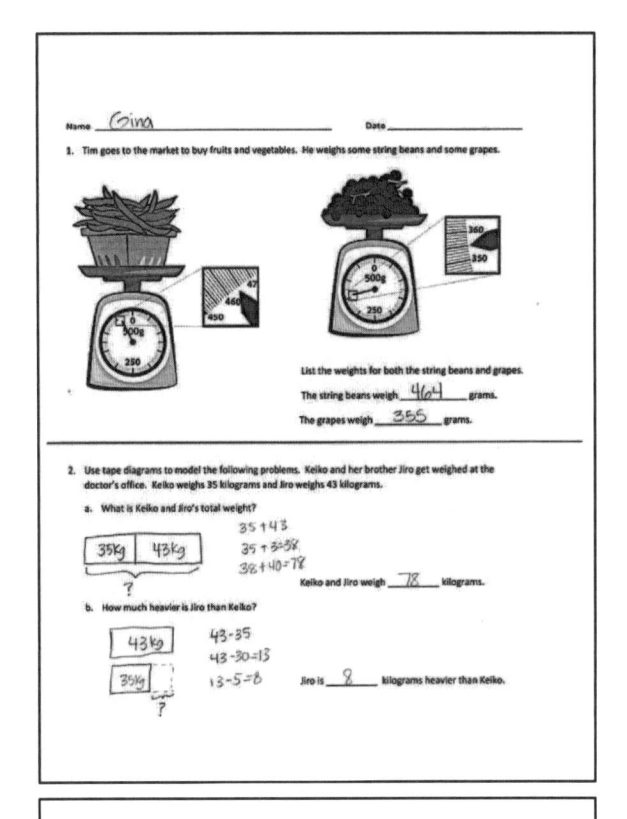

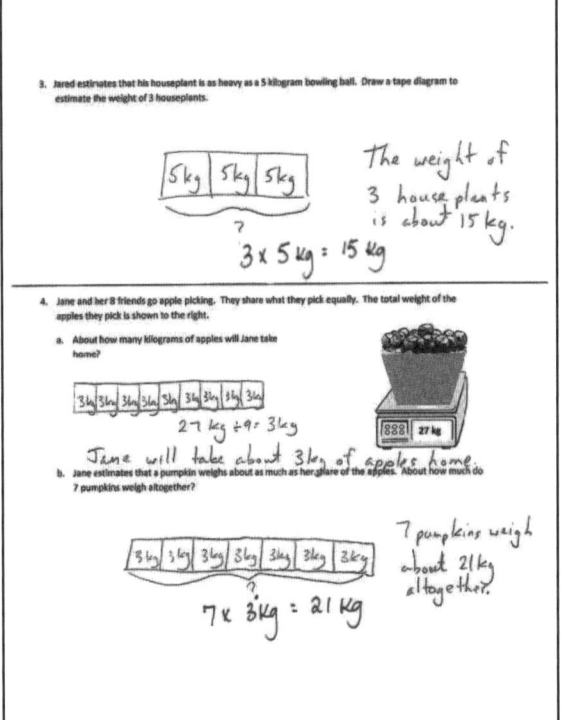

Lesson 8: Solve one-step word problems involving metric weights within 100 and estimate to reason about solutions.

A STORY OF UNITS

Lesson 8 Problem Set 3•2

Name _____ Date _____

1. Tim goes to the market to buy fruits and vegetables. He weighs some string beans and some grapes.

   List the weights for both the string beans and grapes.

   The string beans weigh _____ grams.

   The grapes weigh _____ grams.

---

2. Use tape diagrams to model the following problems. Keiko and her brother Jiro get weighed at the doctor's office. Keiko weighs 35 kilograms, and Jiro weighs 43 kilograms.

   a. What is Keiko and Jiro's total weight?

   Keiko and Jiro weigh _____ kilograms.

   b. How much heavier is Jiro than Keiko?

   Jiro is _____ kilograms heavier than Keiko.

Lesson 8: Solve one-step word problems involving metric weights within 100 and estimate to reason about solutions.

EUREKA MATH

3. Jared estimates that his houseplant is as heavy as a 5-kilogram bowling ball. Draw a tape diagram to estimate the weight of 3 houseplants.

4. Jane and her 8 friends go apple picking. They share what they pick equally. The total weight of the apples they pick is shown to the right.

   a. About how many kilograms of apples will Jane take home?

   b. Jane estimates that a pumpkin weighs about as much as her share of the apples. About how much do 7 pumpkins weigh altogether?

Name _____ Date _____

The weights of a backpack and suitcase are shown below.

7 kg   21 kg

a. How much heavier is the suitcase than the backpack?

b. What is the total weight of 4 identical backpacks?

c. How many backpacks weigh the same as one suitcase?

Name _____ Date _____

1. The weights of 3 fruit baskets are shown below.

   Basket A   Basket B   Basket C
   12 kg   8 kg   16 kg

   a. Basket _____ is the heaviest.

   b. Basket _____ is the lightest.

   c. Basket A is _____ kilograms heavier than Basket B.

   d. What is the total weight of all three baskets?

2. Each journal weighs about 280 grams. What is total weight of 3 journals?

3. Ms. Rios buys 453 grams of strawberries. She has 23 grams left after making smoothies. How many grams of strawberries did she use?

4. Andrea's dad is 57 kilograms heavier than Andrea. Andrea weighs 34 kilograms.

   a. How much does Andrea's dad weigh?

   b. How much do Andrea and her dad weigh in total?

5. Jennifer's grandmother buys carrots at the farm stand. She and her 3 grandchildren equally share the carrots. The total weight of the carrots she buys is shown below.

   a. How many kilograms of carrots will Jennifer get?

   b. Jennifer uses 2 kilograms of carrots to bake muffins. How many kilograms of carrots does she have left?

Lesson 9

Objective: Decompose a liter to reason about the size of 1 liter, 100 milliliters, 10 milliliters, and 1 milliliter.

Suggested Lesson Structure

- Fluency Practice   (4 minutes)
- Concept Development   (46 minutes)
- Student Debrief   (10 minutes)
- **Total Time**   **(60 minutes)**

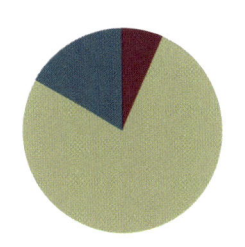

Fluency Practice (4 minutes)

- Decompose 1 Kilogram **3.MD.2**   (4 minutes)

Decompose 1 Kilogram (4 minutes)

Materials: (S) Personal white board

Note: Decomposing 1 kilogram using a number bond helps students relate part–whole thinking to measurement concepts.

- T: (Project a number bond with 1 kg written as the whole.) There are 1,000 grams in 1 kilogram.
- T: (Write 900 g as one of the parts.) On your personal white board, write a number bond by filling in the unknown part.
- S: (Students draw number bond with 100 g, completing the unknown part.)

Continue with the following possible sequence: 500 g, 700 g, 400 g, 600 g, 300 g, 750 g, 650 g, 350 g, 250 g, 850 g, and 150 g.

Concept Development (46 minutes)

Materials: (T) Beaker, 2-liter bottle (empty, top cut off, without label), ten-frame, 12 clear plastic cups (labeled A–L), dropper, one each of the following sizes of containers: cup, pint, quart, gallon (labeled 1, 2, 3, and 4, respectively) (S) Problem Set

---

A NOTE ON STANDARDS ALIGNMENT:

In this lesson, students decompose 1 liter into milliliters following the same procedure used to decompose 1 kilogram into grams used in Lesson 6.

They make connections between metric units and the base ten place value system. The opportunity to make these connections comes from introducing milliliters, which the standards do not include until Grade 4 (**4.MD.1**). Although milliliters are used in Module 2, they are not assessed.

NOTES ON MATERIALS:

Maximize Part 1 by choosing odd-shaped containers, or ones that appear to hold less, for the quart and gallon comparisons. This will challenge students' sense of conservation; they will likely predict that the shorter, wider container holds less than the bottle. Take the opportunity for discussion. How might a shampoo bottle fool you into thinking you are getting more for your money?

A STORY OF UNITS　　　　　　　　　　　　　　　　　　　　　　　　　　　　　　　　　　　Lesson 9　3•2

Part 1: Compare the capacities of containers with different shapes and sizes.

T: (Measure 1 liter of water using a beaker. Pour it into the 2-liter bottle. Use a marker to draw a line at the water level in the bottle, and label it *1 L*. Have containers 1–4 ready.)

T: Which holds more water, a swimming pool or a glass?

S: A swimming pool!

T: Which holds more water, a swimming pool or a bathtub?

S: A swimming pool!

T: Which holds the least amount of water, a swimming pool, a bathtub, or a glass?

S: A glass holds the least amount of water.

T: The amount of liquid a container holds is called its **capacity**. The glass has the smallest capacity because it holds the least amount of water. (Show bottle.) Is this container filled to capacity?

S: No!

T: The amount of water inside measures 1 **liter**. A liter is a unit we use to measure amounts of liquid. To abbreviate the word *liter*, use a capital *L*. (Show the side of the bottle.) Use your finger to write the abbreviation in the air.

T: Let's compare the capacities of different containers by pouring 1 liter into them to see how it fits. (Show Container 1 and the bottle side by side.) Talk to your partner. Predict whether Container 1 holds more than, less than, or about the same as 1 liter. Circle your prediction on Part 1, Problem A of your Problem Set.

S: (Discuss and circle predictions.)

T: I'll pour water from the bottle into Container A to confirm our predictions. (Pour.) Is the capacity of Container 1 more than or less than 1 liter?

S: Less than 1 liter!

T: Does that match your prediction? What surprised you? Why?

S: (Discuss.)

T: Next to the word *actual* on Problem A, write *less*.

Repeat the process with Containers 2–4. (Container 2 holds less than 1 liter, Container 3 holds about the same as 1 liter, and Container 4 holds more than 1 liter.) Then have students complete Problem B.

Part 2: Decompose 1 liter.

T: (Arrange empty cups A–J on the ten-frame, as shown on the next page. Measure and label the water level on Cup K at 100 milliliters and on Cup L at 10 milliliters.)

T: We just compared capacities using a **liquid volume** of 1 liter. We call an amount of liquid *liquid volume*. Whisper the words *liquid volume*.

S: Liquid volume.

T: Now, we're going to decompose 1 liter into smaller units called **milliliters**. Say the word *milliliter*.

S: (Say the word.)

T: To abbreviate *milliliter* we write *mL*. (Model.) Write the abbreviation in the air.

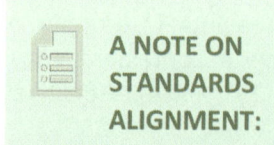

> **A NOTE ON STANDARDS ALIGNMENT:**
>
> The standards do not introduce milliliters until Grade 4 (**4.MD.1**).

T: We'll decompose our liter into 10 equal parts. Each square of our ten-frame shows 1 part. (Show Cup K.) This cup is marked at 100 milliliters. We'll use it to measure the liquid volume that goes into each cup on the ten-frame.

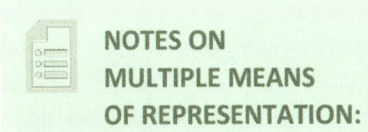

Labeled Cups A–J on a Ten-Frame.

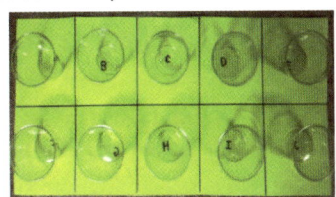

NOTES ON MULTIPLE MEANS OF REPRESENTATION:

Support students in differentiating between the meanings of *capacity* and *liquid volume*. *Capacity* refers to a container and how much the container holds. *Liquid volume* refers to the amount of liquid itself.

T: (Use water from the bottle marked 1 L to fill Cup K to the 100 mL mark. Empty Cup K into Cup A.) How much water is in Cup A?
S: 100 milliliters!
T: (Repeat with Cups B–J.) How many cups are filled with 100 milliliters?
S: 10 cups!
T: Is there any water left in the bottle?
S: No!
T: We partitioned 1 liter of water into 10 parts, each with a liquid volume of about 100 milliliters. Skip-count hundreds to find the total milliliters on the ten-frame. (Point to each cup as students count.)
S: 100, 200, 300, 400, 500, 600, 700, 800, 900, 1,000.
T: How many milliliters of water are in 1 liter?
S: 1,000 milliliters!
T: Talk to your partner about how this equation describes our work. (Write: 1,000 mL ÷ 10 = 100 mL.)

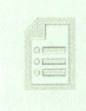

NOTES ON MULTIPLE MEANS OF ENGAGEMENT:

While decomposing, pause to ask students to estimate whether *more than half, half,* or *less than half* of the liter has been poured. Encourage them to reason about estimations using the number of cups on the ten-frame.

S: (Discuss.)
T: Answer Problem C on your Problem Set. Include the equation written on the board.
S: (Skip-count as 9 cups are emptied back into the bottle. Empty the final cup into Cup K.)
T: Let's decompose again. This time we'll pour the 100 milliliters in Cup K into 10 equal parts. (Show Cup L.) This cup is marked at 10 milliliters. We'll use it to measure the liquid volume that goes into each cup on the ten-frame. How many milliliters will be in each of the 10 cups?
S: 10 milliliters. 10 groups of 10 make 100.
T: Cup L is marked at 10 milliliters. (Show Cups K and L side by side.) How do the marks on each cup compare?
S: The mark on Cup L is closer to the bottom.
T: Why is Cup L's mark lower than Cup K's?
S: Cup L shows 10 milliliters. That is less than 100 milliliters. → Cup L shows a smaller liquid volume.
T: (Repeat the process of decomposing as outlined above.)

Lesson 9: Decompose a liter to reason about the size of 1 liter, 100 milliliters, 10 milliliters, and 1 milliliter.

A STORY OF UNITS

Lesson 9 3•2

T: What number sentence represents dividing 100 milliliters into 10 parts?

S: 100 ÷ 10 = 10. → 100 mL ÷ 10 = 10 mL.

T: (Write the equation using units.) Complete Problem D on your Problem Set. Include the equation.

S: (Skip-count as 9 cups are emptied back into the bottle. Empty the final cup into Cup L. Repeat the process used for decomposing 100 milliliters into 10 milliliters, using a dropper to decompose 10 milliliters into cups of 1 milliliter.)

T: How many droppers full of water would it take to fill an entire liter of water?

S: 1,000 droppers full!

T: Answer Problem E. Include the equation.

Measure 1 mL with dropper.

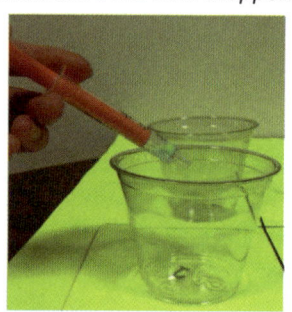

Problem Set (10 minutes)

Students should only need to complete Problems F and G. You may choose to work through these problems as a class, have students work in pairs, or have students work individually. Students should do their personal best to complete the remaining problems within the allotted 10 minutes.

Student Debrief (10 minutes)

Lesson Objective: Decompose a liter to reason about the size of 1 liter, 100 milliliters, 10 milliliters, and 1 milliliter.

The Student Debrief is intended to invite reflection and active processing of the total lesson experience.

Invite students to review their solutions for the problem set. They should check work by comparing answers with a partner before going over answers as a class. Look for misconceptions or misunderstandings that can be addressed in the Debrief. Guide students in a conversation to debrief the Problem Set and process the lesson.

Lesson 9: Decompose a liter to reason about the size of 1 liter, 100 milliliters, 10 milliliters, and 1 milliliter.

A STORY OF UNITS — Lesson 9 3•2

Any combination of the questions below may be used to lead the discussion.

- Revisit predictions from Part 1. Lead a discussion about why students may have thought taller containers had larger **capacities**. Guide students to articulate understanding about conservation and capacity.
- Review the difference between capacity and **liquid volume**.
- In the equations for Part 2, why are the first number and quotient in each followed by the word **milliliters**? Why not the 10?
- How is decomposing 1 **liter** similar to decomposing 1 kilogram?
- How do our decompositions of 1 liter and 1 kilogram remind you of the place value chart?

Exit Ticket (3 minutes)

After the Student Debrief, instruct students to complete the Exit Ticket. A review of their work will help with assessing students' understanding of the concepts that were presented in today's lesson and planning more effectively for future lessons. The questions may be read aloud to the students.

Lesson 9: Decompose a liter to reason about the size of 1 liter, 100 milliliters, 10 milliliters, and 1 milliliter.

Name _____ Date _____

Part 1

a. Predict whether each container holds less than, more than, or about the same as 1 liter.

Container 1 holds   less than / more than / about the same as   1 liter.   Actual:

Container 2 holds   less than / more than / about the same as   1 liter.   Actual:

Container 3 holds   less than / more than / about the same as   1 liter.   Actual:

Container 4 holds   less than / more than / about the same as   1 liter.   Actual:

b. After measuring, what surprised you? Why?

Part 2

c. Illustrate and describe the process of decomposing 1 liter of water into 10 smaller units.

d. Illustrate and describe the process of decomposing Cup K into 10 smaller units.

e. Illustrate and describe the process of decomposing Cup L into 10 smaller units.

f. What is the same about decomposing 1 liter into milliliters and decomposing 1 kilogram into grams?

g. One liter of water weighs 1 kilogram. How much does 1 milliliter of water weigh? Explain how you know.

A STORY OF UNITS Lesson 9 Exit Ticket 3•2

Name _____   Date _____

1. Morgan fills a 1-liter jar with water from the pond. She uses a 100-milliliter cup to scoop water out of the pond and pour it into the jar. How many times will Morgan scoop water from the pond to fill the jar?

2. How many groups of 10 milliliters are in 1 liter? Explain.

There are _____ groups of 10 milliliters in 1 liter.

Name _____ Date _____

1. Find containers at home that have a capacity of about 1 liter. Use the labels on containers to help you identify them.

   a.

Name of Container
Example: Carton of orange juice

   b. Sketch the containers. How do their sizes and shapes compare?

2. The doctor prescribes Mrs. Larson 5 milliliters of medicine each day for 3 days. How many milliliters of medicine will she take altogether?

3. Mrs. Goldstein pours 3 juice boxes into a bowl to make punch. Each juice box holds 236 milliliters. How much juice does Mrs. Goldstein pour into the bowl?

4. Daniel's fish tank holds 24 liters of water. He uses a 4-liter bucket to fill the tank. How many buckets of water are needed to fill the tank?

5. Sheila buys 15 liters of paint to paint her house. She pours the paint equally into 3 buckets. How many liters of paint are in each bucket?

A STORY OF UNITS

Lesson 10 3•2

Lesson 10

Objective: Estimate and measure liquid volume in liters and milliliters using the vertical number line.

Suggested Lesson Structure

- ■ Fluency Practice   (10 minutes)
- ■ Application Problem   (5 minutes)
- ■ Concept Development   (35 minutes)
- ■ Student Debrief   (10 minutes)
-   **Total Time**   **(60 minutes)**

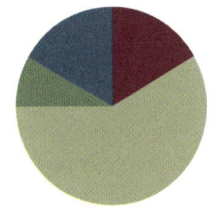

Fluency Practice (10 minutes)

- Milliliter Counting **3.MD.2**   (2 minutes)
- Decompose 1 Liter **3.MD.2**   (4 minutes)
- Group Counting **3.OA.1**   (4 minutes)

Milliliter Counting (2 minutes)

Note: This activity reviews Lesson 9 and lays the foundation for eventually composing compound units of liters and milliliters in Grade 4.

T: There are 1,000 milliliters in 1 liter. Count by 100 milliliters to 1 liter.

S: 100 milliliters, 200 milliliters, 300 milliliters, 400 milliliters, 500 milliliters, 600 milliliters, 700 milliliters, 800 milliliters, 900 milliliters, 1 liter.

Decompose 1 Liter (4 minutes)

Materials: (S) Personal white board

Note: Decomposing 1 liter using a number bond helps students relate part–whole thinking to measurement concepts.

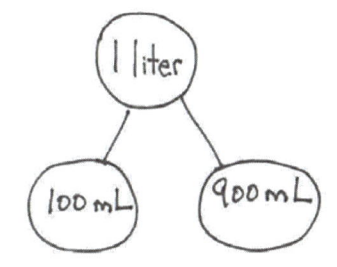

T: (Project a number bond with *1 liter* written as the whole.) There are 1,000 milliliters in 1 liter.

T: (Write 900 mL as one of the parts.) On your personal white board, write a number bond by filling in the unknown part.

S: (Draw number bond with 100 mL, completing the unknown part.)

Continue with possible sequence of 500 mL, 700 mL, 400 mL, 600 mL, 300 mL, 750 mL, 650 mL, 350 mL, 250 mL, 850 mL, and 150 mL.

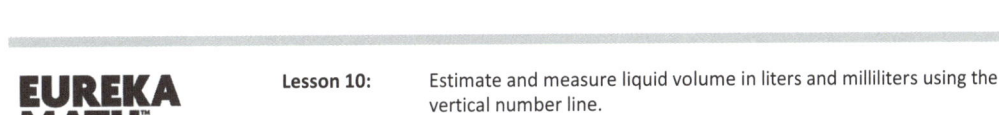

Lesson 10: Estimate and measure liquid volume in liters and milliliters using the vertical number line.

Group Counting (4 minutes)

Note: Group counting reviews interpreting multiplication as repeated addition. It reviews foundational strategies for multiplication from Module 1 and anticipates Module 3.

Direct students to count forward and backward, occasionally changing the direction of the count:

- Threes to 30
- Fours to 40
- Sixes to 60
- Sevens to 70
- Eights to 80
- Nines to 90

As students' fluency with skip-counting increases, have them track the number of groups counted with their fingers in order to make the connection to multiplication.

Application Problem (5 minutes)

Subha drinks 4 large glasses of water each day. How many large glasses of water does she drink in 7 days?

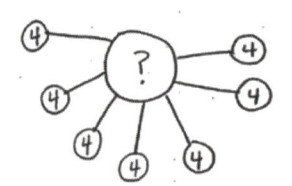

7 x 4 glasses = 28 glasses.

Subha drinks 28 glasses of water.

Note: This problem activates prior knowledge about solving multiplication word problems using units of 4. It is designed to lead into a discussion about liquid volume in the Concept Development.

Concept Development (35 minutes)

Materials: (T) 1-liter beaker (S) Pitcher of water (1 per group), empty 2-liter bottle with top cut off (1 per group), 1 plastic cup pre-measured and labeled at 100 mL, 1 permanent marker, Problem Set

NOTES ON MATERIALS:

The bottles used in this exploration should be as close to the shape of a cylinder as possible. This will create a more precise vertical number line with tick marks that are equidistant from one another. Many soda bottles have grooves on the bottom and a thinner waistline, which will skew the tick marks and create uneven intervals on the number line.

In this lesson, the 1-liter beaker is used to validate the ones students construct. If access to a 1-liter beaker is not available, show a photo instead.

A STORY OF UNITS

Lesson 10 3•2

Part 1: Create a vertical number line marked at 100 mL intervals.

Empty bottle with number line

T: (Make groups of three students.) Each group will measure liquid volume to make a measuring bottle that contains 1 liter of water, similar to the one we used yesterday. Each group member has a job. One person will be the measurer, one will be the pourer, and the other will be the marker. Take 30 seconds to decide on jobs.

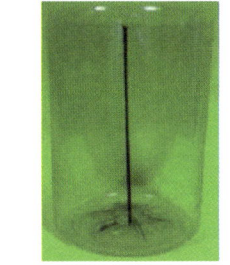

S: (Decide.)

T: The marker should draw a straight, vertical line from top to bottom (pictured on the right). These are the rest of the directions:

- The measurer measures 100 milliliters of water by pouring from the pitcher into the plastic cup.
- The pourer holds the plastic cup in place and helps the measurer know when to stop. Then the pourer pours the water from the cup into the bottle.
- The marker makes horizontal lines to show each new water level on the side of the bottle. Each horizontal line should cross the vertical line. The horizontal lines should be about the same size, and one should be right above the other.

T: There are 1,000 milliliters in 1 liter of water. You are measuring 100 milliliters each time. Think back to yesterday. How many times will you need to measure and mark 100 milliliters of water to make 1 liter?

S: 10 times.

T: Go ahead and get started.

S: (Measure, pour, and mark until there are 10 horizontal lines on the bottle and 1 liter of water inside.)

T: What do the tick marks and line remind you of?

S: They look like the number line! → It's going up and down instead of sideways.

MP.6

T: Another way to say up-and-down is *vertical*. It's a vertical number line. Point to the tick mark that shows the *most* liquid volume.

S: (Point to the top-most horizontal mark.)

T: Use the word *milliliters* or *liters* to tell your group the capacity indicated by that mark.

S: 1,000 milliliters. → 1 liter.

Labels at 100 mL, 500 mL, 1 L

T: To the right of the mark, label 1 L.

Repeat the process for the mark that shows the *least* liquid volume and label 100 mL.

T: With your group, use the vertical number line to find the mark that shows about **halfway** to 1 liter. Discuss the value of the mark in milliliters. Make sure you all agree.

S: (Find the mark; agree that the value is 500 mL.)

T: Label the halfway mark.

S: (Label 500 mL.)

T: You've made a tool that scientists and mathematicians use to measure liquid volume. It's called a beaker. (Show a beaker.) Work with your group to answer all three parts of Problem 1 on your Problem Set.

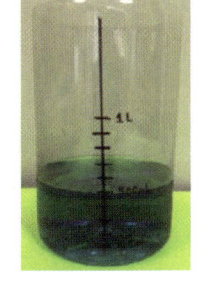

Lesson 10: Estimate and measure liquid volume in liters and milliliters using the vertical number line.

EUREKA MATH

Part 2: Use the vertical number line to estimate and precisely measure liquid volume.

S: (Groups pour the liter of water from the measuring bottle into the pitcher.)

T: A small water bottle has about 200 milliliters of water inside. Let's see what 200 milliliters looks like. Pour from your pitcher to the measuring bottle to see the capacity of a small water bottle.

S: (Pour and measure 200 mL.)

T: How did your group use the vertical number line to measure?

S: Each tick mark represents 100 milliliters. We knew the water level was at 200 milliliters when it reached the second tick mark.

T: Is the water level in your bottle less than halfway, more than halfway, or about halfway to a liter?

S: Less than halfway.

T: A larger water bottle has about 500 milliliters of water inside. How many milliliters should you add to your measuring bottle so that the liquid volume is the same as that of a larger water bottle?

S: 300 milliliters.

T: How many tick marks higher should the water level rise if you are adding 300 milliliters?

S: Three tick marks higher.

T: Add 300 milliliters of water to your measuring bottle.

S: (Pour and measure 300 milliliters.)

T: Is the water level in your bottle less than halfway, more than halfway, or about halfway to a liter?

S: About halfway.

> **NOTES ON MULTIPLE MEANS OF ENGAGEMENT:**
>
> Encourage students to use precise strategies rather than estimating half by sight. For example, they might skip-count or divide 10 marks by 2.

> **NOTES ON MULTIPLE MEANS OF ENGAGEMENT:**
>
> If appropriate, increase the level of difficulty by asking students to estimate *less than halfway* between tick marks and *more than halfway* using the following examples:
> - 225 mL, 510 mL (less than)
> - 675 mL, 790 mL (more than)

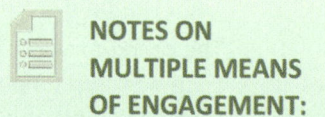

Repeat the process with the following sequence:

- 700 mL, 900 mL, 1,000 mL
- 250 mL, 450 mL
  (These will be estimates. This is an opportunity to discuss *halfway between* two tick marks.)

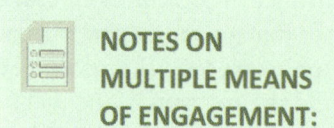

Lesson 10 3•2

Problem Set (10 minutes)

Students should do their best to complete Problems 2–4 within the allotted 10 minutes. For some classes, it may be appropriate to modify the assignment by specifying which problems they work on first.

Student Debrief (10 minutes)

Lesson Objective: Estimate and measure liquid volume in liters and milliliters using the vertical number line.

The Student Debrief is intended to invite reflection and active processing of the total lesson experience.

Invite students to review their solutions for the Problem Set. They should check work by comparing answers with a partner before going over answers as a class. Look for misconceptions or misunderstandings that can be addressed in the Debrief. Guide students in a conversation to debrief the Problem Set and process the lesson.

Any combination of the questions below may be used to lead the discussion.

- In Problem 4, describe how the position of the points plotted in Part (a) helped you solve Parts (b) and (c).
- Students may have different answers for Problem 4(d). (Barrel B is closest to 70, but Barrel A has enough capacity to hold 70 liters, plus a little extra.) Invite students with both answers to explain their thinking.
- Compare the beaker with your measuring bottle.
- How would we have labeled our vertical number lines differently if we had measured 10 mL instead of 100 mL cups to make our measuring bottles?
- If we had measured 10 mL instead of 100 mL cups to make our measuring bottles, would our **halfway** mark have been the same or different? How do you know?
- Would our estimates change if our bottles had marks at every 10 mL instead of every 100 mL?

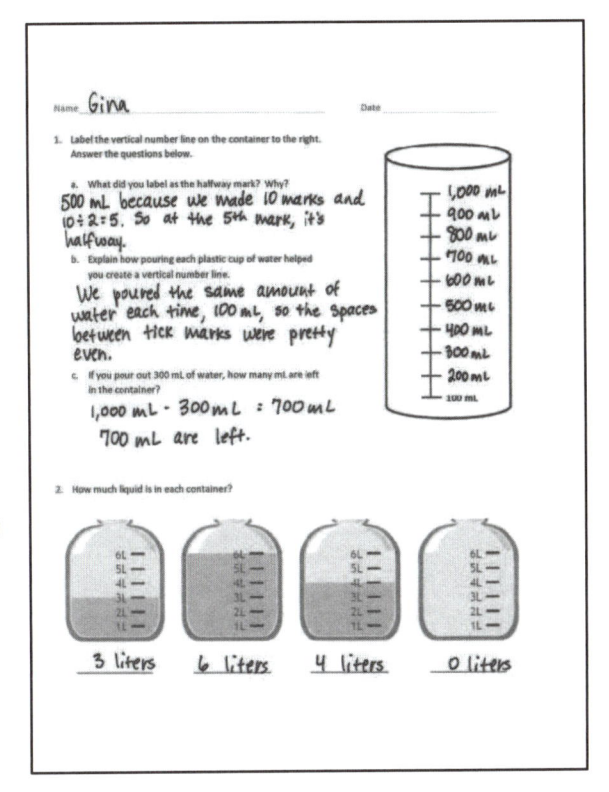

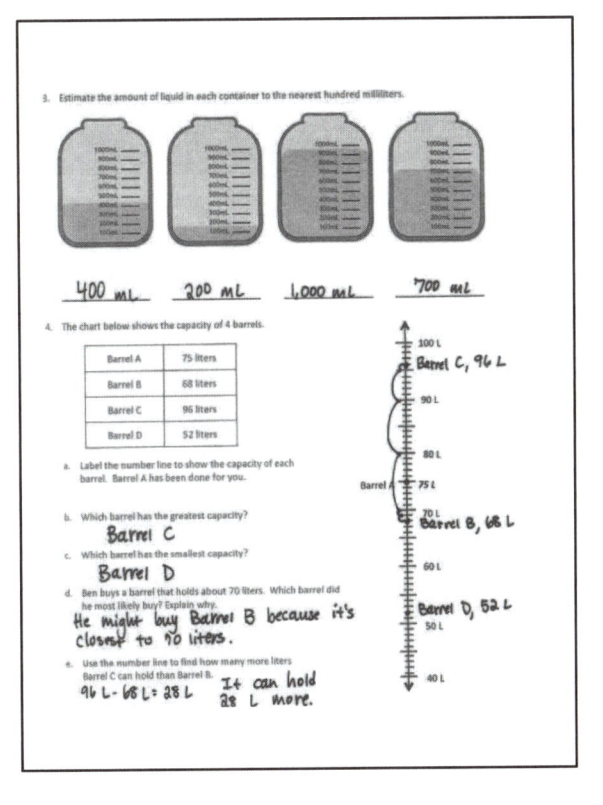

Exit Ticket (3 minutes)

After the Student Debrief, instruct students to complete the Exit Ticket. A review of their work will help with assessing students' understanding of the concepts that were presented in today's lesson and planning more effectively for future lessons. The questions may be read aloud to the students.

A STORY OF UNITS

Lesson 10 Problem Set 3•2

Name_____   Date _____

1. Label the vertical number line on the container to the right. Answer the questions below.

   a. What did you label as the halfway mark? Why?

   b. Explain how pouring each plastic cup of water helped you create a vertical number line.

   c. If you pour out 300 mL of water, how many mL are left in the container?

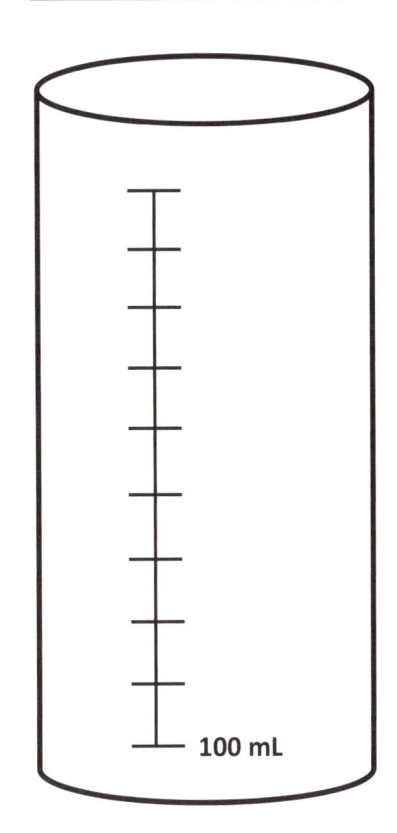

   100 mL

2. How much liquid is in each container?

   6L 5L 4L 3L 2L 1L

   _____   _____   _____   _____

Lesson 10: Estimate and measure liquid volume in liters and milliliters using the vertical number line.

A STORY OF UNITS

Lesson 10 Problem Set 3•2

3. Estimate the amount of liquid in each container to the nearest hundred milliliters.

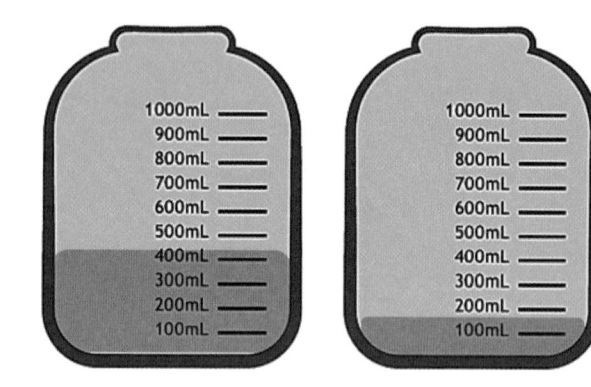

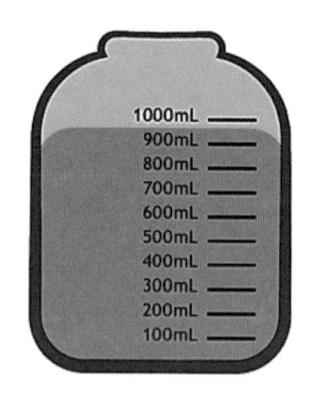

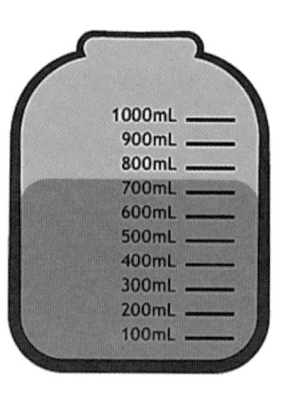

   _____   _____   _____   _____

4. The chart below shows the capacity of 4 barrels.

Barrel A	75 liters
Barrel B	68 liters
Barrel C	96 liters
Barrel D	52 liters

   a. Label the number line to show the capacity of each barrel. Barrel A has been done for you.

   b. Which barrel has the greatest capacity?

   c. Which barrel has the smallest capacity?

   d. Ben buys a barrel that holds about 70 liters. Which barrel did he most likely buy? Explain why.

   e. Use the number line to find how many more liters Barrel C can hold than Barrel B.

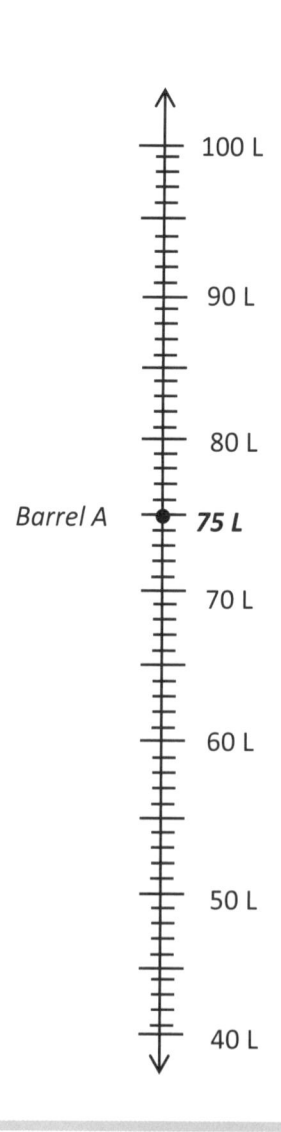

   Barrel A — 75 L

122   Lesson 10:   Estimate and measure liquid volume in liters and milliliters using the vertical number line.

Name _____   Date _____

1. Use the number line to record the capacity of the containers.

Container	Capacity in Liters
A	
B	
C	

2. What is the difference between the capacity of Container A and Container C?

Name_____ Date_____

1. How much liquid is in each container?

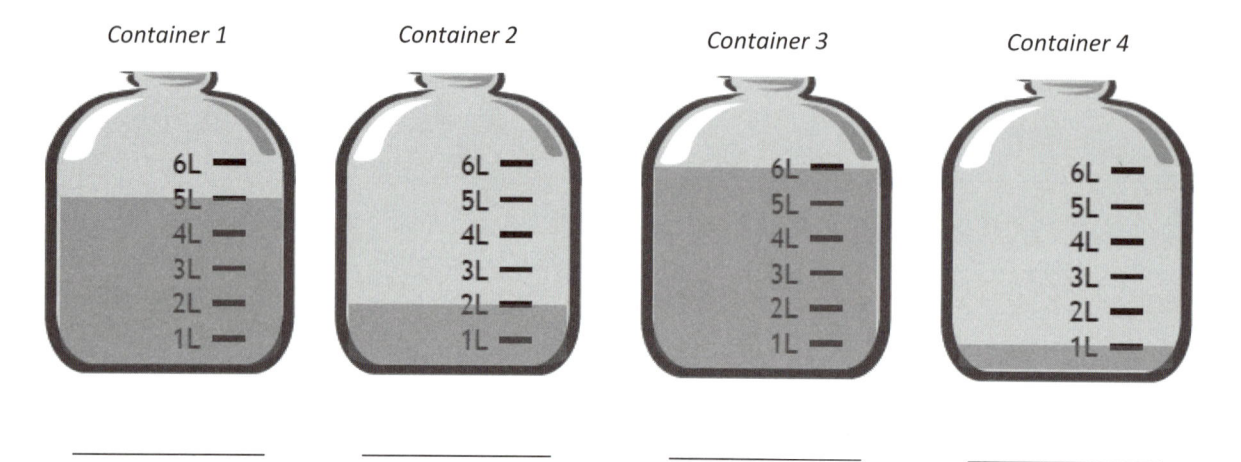

   Container 1: _____ Container 2: _____ Container 3: _____ Container 4: _____

2. Jon pours the contents of Container 1 and Container 3 above into an empty bucket. How much liquid is in the bucket after he pours the liquid?

3. Estimate the amount of liquid in each container to the nearest liter.

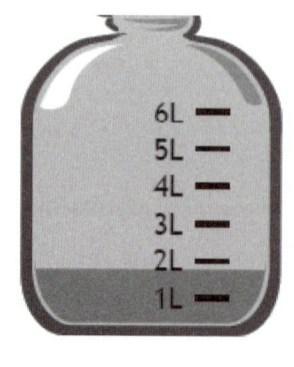

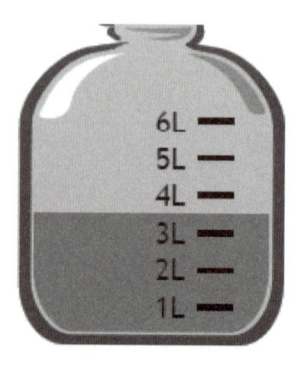

   _____ _____ _____ _____

Lesson 10: Estimate and measure liquid volume in liters and milliliters using the vertical number line.

4. Kristen is comparing the capacity of gas tanks in different size cars. Use the chart below to answer the questions.

Size of Car	Capacity in Liters
Large	74
Medium	57
Small	42

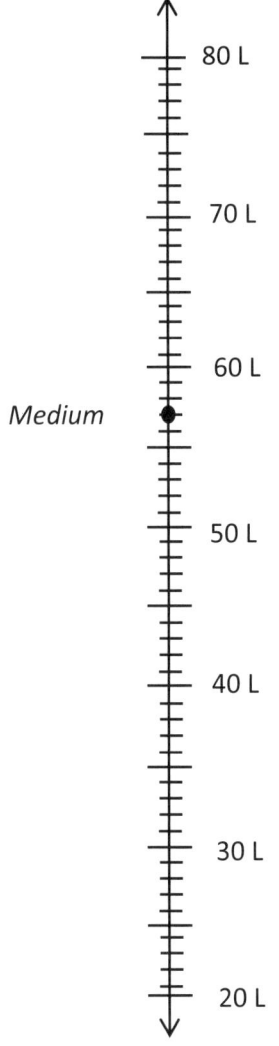

a. Label the number line to show the capacity of each gas tank. The medium car has been done for you.

b. Which car's gas tank has the greatest capacity?

c. Which car's gas tank has the smallest capacity?

d. Kristen's car has a gas tank capacity of about 60 liters. Which car from the chart has about the same capacity as Kristen's car?

e. Use the number line to find how many more liters the large car's tank holds than the small car's tank.

A STORY OF UNITS

Lesson 11 3•2

Lesson 11

Objective: Solve mixed word problems involving all four operations with grams, kilograms, liters, and milliliters given in the same units.

Suggested Lesson Structure

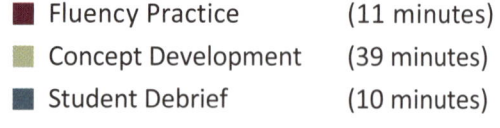

■ Fluency Practice   (11 minutes)
■ Concept Development   (39 minutes)
■ Student Debrief   (10 minutes)
   Total Time   **(60 minutes)**

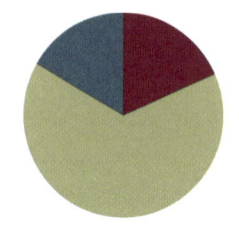

Fluency Practice (11 minutes)

- Rename Tens **3.NBT.3**   (3 minutes)
- Halfway on the Number Line **3.NBT.1**   (4 minutes)
- Read a Beaker **3.MD.1**   (4 minutes)

Rename Tens (3 minutes)

Materials: (T) Place value cards (S) Personal white board

Note: This activity anticipates rounding in the next topic. If necessary, use place value cards to quickly review place value with students.

    T: (Write 7 tens = ____.) Say the number.

    S: 70.

Continue with the following possible sequence: 8 tens, 9 tens, and 10 tens.

    T: (Write 11 tens = ____.) On your personal white board, fill in the number sentence.

    S: (Write 11 tens = 110.)

Continue with the following possible sequence: 12 tens, 16 tens, 19 tens, and 15 tens.

Place Value Cards

Halfway on the Number Line (4 minutes)

Materials: (S) Personal white board

Note: This activity anticipates rounding in the next topic. Practicing this skill in isolation lays a foundation for conceptually understanding rounding on a vertical number line.

- T: (Project a vertical line with endpoints labeled 0 and 10.) What's halfway between 0 tens and 1 ten?
- S: 5.
- T: (Write 5 halfway between 0 and 10.)

Repeat process with endpoints labeled 10 and 20.

- T: Draw a vertical number line on your board. Make tick marks at each end and one for the halfway point.
- S: (Draw number line.)
- T: (Write 3 tens and 4 tens.) Label the tick marks at each end and at the halfway point.
- S: (Label 30 as the bottom point, 40 as the top point, and 35 as the halfway point.)

Continue with the following possible sequence: 60 and 70, 80 and 90, 40 and 50, and 50 and 60.

Read a Beaker (4 minutes)

Materials: (T) Beaker images (S) Personal white board

Note: This activity reviews Lesson 10.

- T: (Show image of a beaker with a capacity of 4 liters.) Start at the bottom of the beaker and count by 1 liter. (Move finger from the bottom to each tick mark as students count.)
- S: 1 liter, 2 liters, 3 liters, 4 liters.
- T: I'll shade in the beaker to show how much water it's holding. Write the liquid volume on your board. (Shade in 1 liter.)
- S: (Write 1 liter.)

Repeat the process, varying the liquid height.

Repeat the process with a beaker partitioned into 10 equal parts, filling in increments of 100 milliliters.

Repeat the process with a beaker partitioned into 2 equal parts, filling in increments of 500 milliliters.

Concept Development (39 minutes)

Materials: (T) Scale (S) Spring scales, digital scales, beakers (mL), personal white board

Problem 1: Solve word problems involving addition and subtraction.

- T: (Project.) A pet mouse weighs 34 grams. A pet hamster weighs 126 grams more than the mouse. How much does the pet hamster weigh? Model the problem on your board.
- S: (Model.)

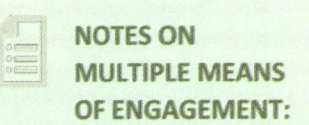

> **NOTES ON MULTIPLE MEANS OF ENGAGEMENT:**
>
> This lesson includes an abundance of word problems given in all four operations. It is unlikely that there will be time for them all. As decisions are made about pacing, select problems involving operations with which the class most needs practice, and intentionally vary the problem types.

A STORY OF UNITS   Lesson 11   3•2

T: Talk with your partner: Is there a simplifying strategy you might use to find how much the hamster weighs?

S: 126 grams is almost 130 grams. We can use the 4 from 34 to complete the ten in 126 and make 130. Then it's just 30 + 130. That's easy!

T: How might this strategy help us solve similar problems using mental math?

S: We can look for other problems with 6 in the ones place and see if getting 4 makes a simpler problem. → We can look for ways to make a ten.

$$34g + 126g = ?$$
$$\wedge$$
$$30 \quad 4$$

$$30 + 4 + 126 =$$
$$30 + 130 = 160g$$

As time allows, repeat the process.

- *Add to with result unknown:* Judith squeezes 140 milliliters of lemon juice to make 1 liter of lemonade. How many milliliters of lemon juice are in 2 liters of lemonade?
- *Take from with change unknown:* Robert's crate of tools weighs 12 kilograms. He takes his power tools out. Now the crate weighs 4 kilograms. How many kilograms do the power tools weigh?

Problem 2: Solve word problems involving multiplication.

T: (Project.) A pitcher of shaved ice needs 5 milliliters of food coloring to turn red. How many milliliters of food coloring are needed to make 9 pitchers of shaved ice red? Explain to your partner how you would model and solve this problem. (Pause.)

T: Go ahead and solve.

S: (Solve problem.)

T: (Pick two students who used different strategies to share.)

S: (Share.)

As time allows, repeat the process.

- *Equal groups with unknown product:* Alyssa drinks 3 liters of water every day. How many liters will she drink in 8 days?
- *Equal groups with unknown product:* There are 4 grams of almonds in each bag of mixed nuts. How many grams of almonds are in 7 bags?

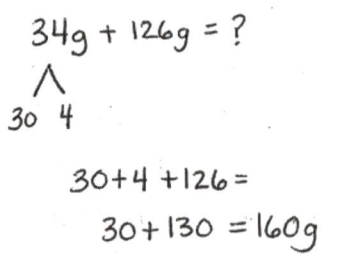

> **NOTES ON MULTIPLE MEANS OF ENGAGEMENT:**
>
> Students may come up with a variety of strategies. Strategically choose students to share their work, highlighting for the rest of the class particularly efficient methods. Use what students share to build a bank of strategies, and encourage students to try a friend's strategy to solve subsequent problems.

> **NOTES ON MULTIPLE MEANS OF ACTION AND EXPRESSION:**
>
> Develop students' sense of liquid volume by having them estimate to model 140 mL and then 280 mL after solving the *add to with change unknown* problem using measuring bottles from Lesson 10.

A STORY OF UNITS   Lesson 11 3•2

Problem 3: Solve word problems involving division.

T: Let's work in groups to solve the following problem. (Group students.)
T: (Project.) At the pet shop there are 36 liters of water in a tank. Each fish bowl holds 4 liters. How many fish bowls can the shopkeeper fill using the water in the tank?
T: Go ahead and solve.
S: (Solve problem.)
T: (Pick groups that used different strategies to share.)
S: (Share.)

As time allows, repeat the process:

- *Equal groups with number of groups unknown*: Every day the school garden gets watered with 7 liters of water. How many days pass until the garden has been watered with 49 liters?
- *Equal groups with group size unknown*: A bin at the grocery store holds 9 kilograms of walnuts. The total value of 9 kilograms of walnuts is $36. How much does 1 kilogram of walnuts cost?

As time allows, have students work in pairs to solve one-step word problems using all four operations.

- *Take apart with addend unknown*: Together an orange and a mango weigh 637 grams. The orange weighs 385 grams. What is the weight of the mango?
- *Compare with difference unknown*: A rabbit weighs 892 grams. A guinea pig weighs 736 grams. How much more does the rabbit weigh than the guinea pig?
- *Equal groups with group size unknown*: Twenty-four kilograms of pineapple are needed to make 4 identical fruit platters. How many kilograms of pineapple are required to make 1 fruit platter?
- *Equal groups with unknown product:* The capacity of a pitcher is 3 liters. What is the capacity of 9 pitchers?
- *Add to with result unknown:* Jack uses a beaker to measure 250 milliliters of water. Angie measures double that amount. How many milliliters of water does Angie measure?

Problem Set (10 minutes)

Students should do their personal best to complete the problem set within the allotted 10 minutes. For some classes, it may be appropriate to modify the assignment by specifying which problems they work on first. Some problems do not specify a method for solving. Students should solve these problems using the RDW approach used for Application Problems.

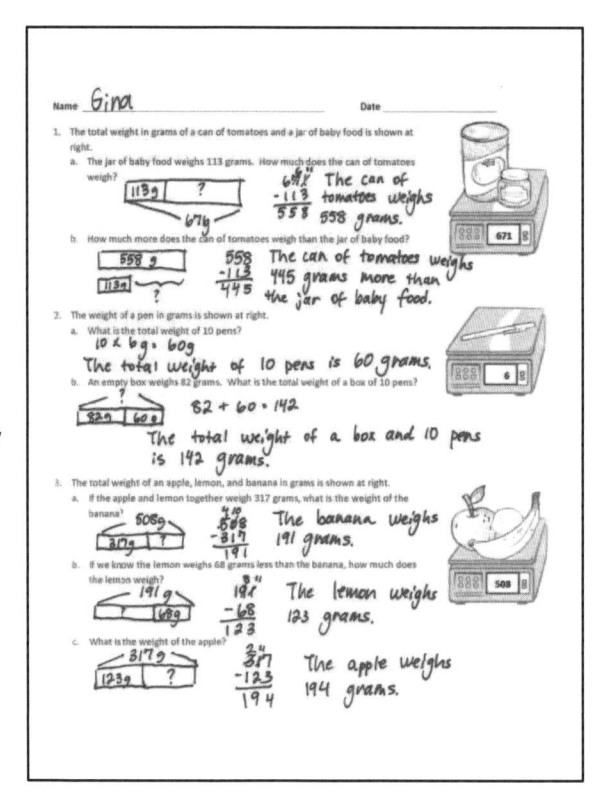

Lesson 11: Solve mixed word problems involving all four operations with grams, kilograms, liters, and milliliters given in the same units.

Student Debrief (10 minutes)

Lesson Objective: Solve mixed word problems involving all four operations with grams, kilograms, liters, and milliliters given in the same units.

The Student Debrief is intended to invite reflection and active processing of the total lesson experience.

Invite students to review their solutions for the Problem Set. They should check work by comparing answers with a partner before going over answers as a class. Look for misconceptions or misunderstandings that can be addressed in the Debrief. Guide students in a conversation to debrief the Problem Set and process the lesson.

Any combination of the questions below may be used to lead the discussion.

- What models did you use to solve the word problems?
- Explain the process you used for solving Problem 1. Did you use a special strategy? What was it?
- **MP.7** What pattern did you notice between Problems 4, 5, and 6? How did that pattern help you solve the problems?
- Explain why Problem 6 was more challenging to solve than Problems 4 and 5.
- Look at Problem 6. Why is it important to measure the capacity of an object before dividing into equal amounts?

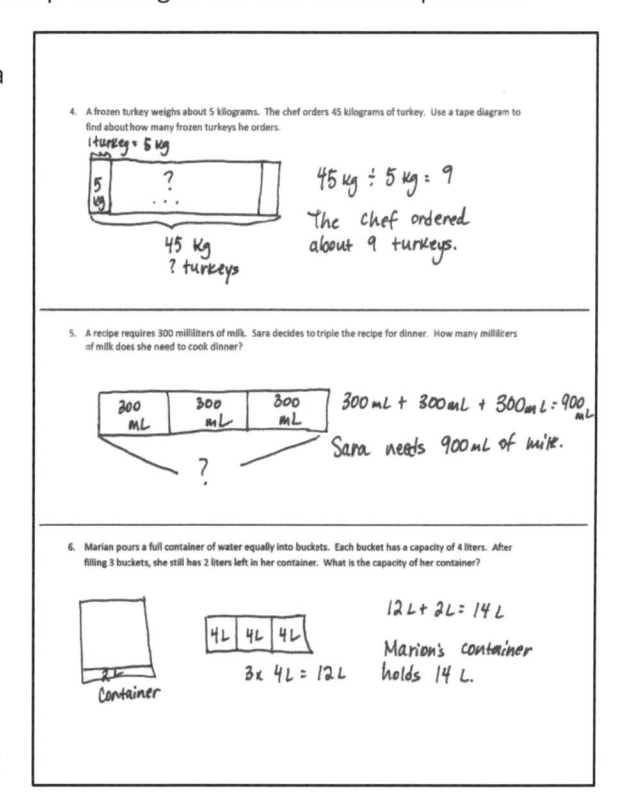

Exit Ticket (3 minutes)

After the Student Debrief, instruct students to complete the Exit Ticket. A review of their work will help with assessing students' understanding of the concepts that were presented in today's lesson and planning more effectively for future lessons. The questions may be read aloud to the students.

Name _____   Date _____

1. The total weight in grams of a can of tomatoes and a jar of baby food is shown to the right.

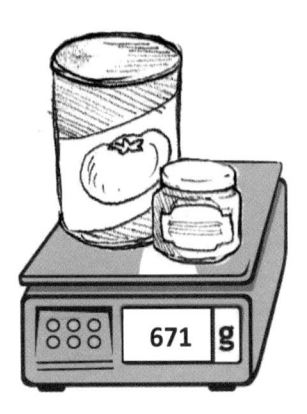

   a. The jar of baby food weighs 113 grams. How much does the can of tomatoes weigh?

   b. How much more does the can of tomatoes weigh than the jar of baby food?

2. The weight of a pen in grams is shown to the right.

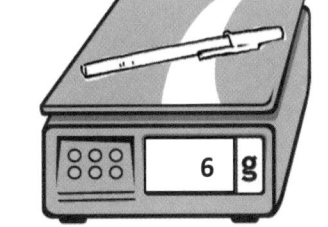

   a. What is the total weight of 10 pens?

   b. An empty box weighs 82 grams. What is the total weight of a box of 10 pens?

3. The total weight of an apple, lemon, and banana in grams is shown to the right.

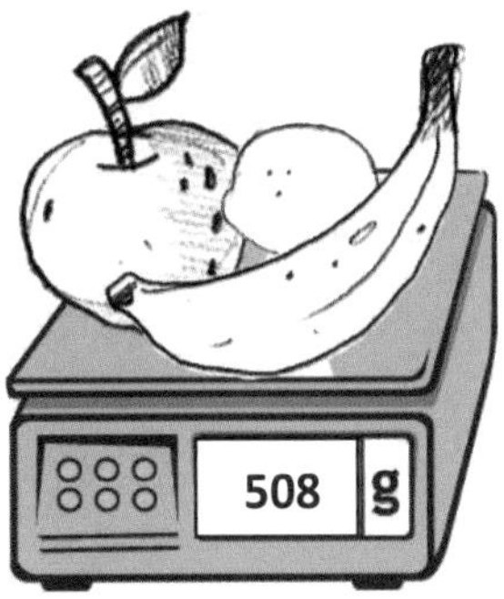

   a. If the apple and lemon together weigh 317 grams, what is the weight of the banana?

   b. If we know the lemon weighs 68 grams less than the banana, how much does the lemon weigh?

   c. What is the weight of the apple?

4. A frozen turkey weighs about 5 kilograms. The chef orders 45 kilograms of turkey. Use a tape diagram to find about how many frozen turkeys he orders.

5. A recipe requires 300 milliliters of milk. Sara decides to triple the recipe for dinner. How many milliliters of milk does she need to cook dinner?

6. Marian pours a full container of water equally into buckets. Each bucket has a capacity of 4 liters. After filling 3 buckets, she still has 2 liters left in her container. What is the capacity of her container?

A STORY OF UNITS Lesson 11 Exit Ticket 3•2

Name _____ Date _____

The capacities of three cups are shown below.

Cup A
160 mL

Cup B
280 mL

Cup C
237 mL

a. Find the total capacity of the three cups.

b. Bill drinks exactly half of Cup B. How many milliliters are left in Cup B?

c. Anna drinks 3 cups of tea from Cup A. How much tea does she drink in total?

A STORY OF UNITS **Lesson 11 Homework 3•2**

Name _____   Date _____

1. Karina goes on a hike. She brings a notebook, a pencil, and a camera. The weight of each item is shown in the chart. What is the total weight of all three items?

Item	Weight
Notebook	312 g
Pencil	10 g
Camera	365 g

The total weight is _____ grams.

2. Together a horse and its rider weigh 729 kilograms. The horse weighs 625 kilograms. How much does the rider weigh?

The rider weighs _____ kilograms.

Lesson 11: Solve mixed word problems involving all four operations with grams, kilograms, liters, and milliliters given in the same units.

3. Theresa's soccer team fills up 6 water coolers before the game. Each water cooler holds 9 liters of water. How many liters of water do they fill?

4. Dwight purchased 48 kilograms of fertilizer for his vegetable garden. He needs 6 kilograms of fertilizer for each bed of vegetables. How many beds of vegetables can he fertilize?

5. Nancy bakes 7 cakes for the school bake sale. Each cake requires 5 milliliters of oil. How many milliliters of oil does she use?

Name _____ Date _____

1. Fatima runs errands.

   a. The clock to the right shows what time she leaves home. What time does she leave?

   Fatima leaves home.

   b. It takes Fatima 17 minutes to go from her home to the market. Use the number line below to show what time she gets to the market.

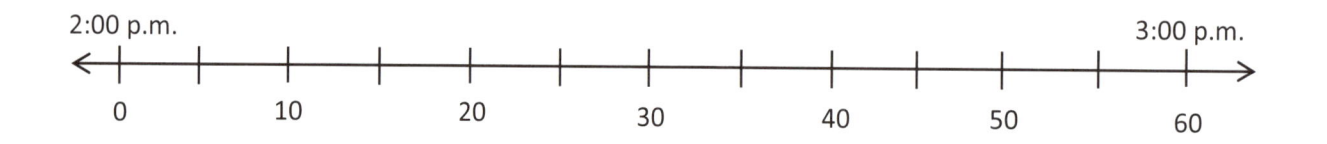

   2:00 p.m.   3:00 p.m.
   |----|----|----|----|----|----|----|----|----|----|----|----|
   0   10   20   30   40   50   60

   c. The clock to the right shows what time Fatima leaves the market. What time does she leave the market?

   Fatima leaves the market.

   d. How long does Fatima spend at the market?

2. At the market, Fatima uses a scale to weigh a bag of almonds and a bag of raisins, shown below. What is the total weight of the almonds and raisins?

3. The amount of juice in 1 bottle is shown to the right. Fatima needs 18 liters for a party. Draw and label a tape diagram to find how many bottles of juice she should buy.

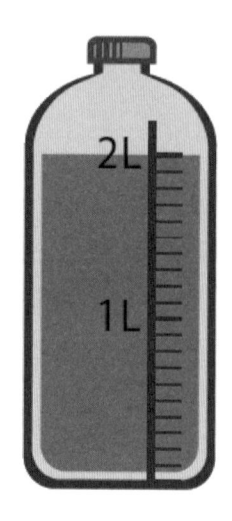

4. Altogether, Fatima's lettuce, broccoli, and peas weigh 968 grams. The total weight of her lettuce and broccoli is shown to the right. Write and solve a number sentence to find how much the peas weigh.

5. Fatima weighs a watermelon, shown to the right.

   a. How much does the watermelon weigh?

   b. Leaving the store Fatima thinks, "Each bag of groceries seems as heavy as a watermelon!" Use Fatima's idea about the weight of the watermelon to estimate the total weight of 7 bags.

   c. The grocer helps carry about 9 kilograms. Fatima carries the rest. Estimate how many kilograms of groceries Fatima carries.

   d. It takes Fatima 12 minutes to drive to the bank after she leaves the store and then 34 more minutes to drive home. How many minutes does Fatima drive after she leaves the store?

Mid-Module Assessment Task 3•2

Mid-Module Assessment Task
Standards Addressed
Topics A–B

Use place value understanding and properties of operations to perform multi-digit arithmetic. (A range of algorithms may be used.)

3.NBT.2 Fluently add and subtract within 1000 using strategies and algorithms based on place value, properties of operations, and/or the relationship between addition and subtraction.

Solve problems involving measurement and estimation of intervals of time, liquid volumes, and masses of objects.

3.MD.1 Tell and write time to the nearest minute and measure time intervals in minutes. Solve word problems involving addition and subtraction of time intervals in minutes, e.g., by representing the problem on a number line diagram.

3.MD.2 Measure and estimate liquid volumes and masses of objects using standard units of grams (g), kilograms (kg), and liters (l). (Excludes compound units such as cm³ and finding the geometric volume of a container.) Add, subtract, multiply, or divide to solve one-step word problems involving masses or volumes that are given in the same units, e.g., by using drawings (such as a beaker with a measurement scale) to represent the problem. (Excludes multiplicative comparison problems, i.e., problems involving notions of "times as many"; see CCLS Glossary, Table 2.)

Evaluating Student Learning Outcomes

A Progression Toward Mastery is provided to describe steps that illuminate the gradually increasing understandings that students develop on their way to proficiency. In this chart, this progress is presented from left (Step 1) to right (Step 4). The learning goal for students is to achieve Step 4 mastery. These steps are meant to help teachers and students identify and celebrate what the students CAN do now and what they need to work on next.

A STORY OF UNITS

Mid-Module Assessment Task 3•2

A Progression Toward Mastery

Assessment Task Item	STEP 1 Little evidence of reasoning without a correct answer. (1 Point)	STEP 2 Evidence of some reasoning without a correct answer. (2 Points)	STEP 3 Evidence of some reasoning with a correct answer or evidence of solid reasoning with an incorrect answer. (3 Points)	STEP 4 Evidence of solid reasoning with a correct answer. (4 Points)
1 3.MD.1	Student gives an incorrect answer. Attempt shows student may not understand the meaning of the questions.	Student gives an incorrect answer with a reasonable attempt: • Accurately reads the clocks. • Attempts to use a number line. • Attempts to calculate Part (d).	Student gives a partially correct answer: • Accurately reads clocks. • Attempts to use a number line. • Accurately calculates Part (d).	Student correctly answers each part of the question: a. Reads 2:07 on the clock. b. Draws a number line to show 2:24. c. Reads 2:53 on the clock. d. Calculates 29 minutes.
2 3.NBT.2 3.MD.2	Student gives an incorrect answer. Attempt shows the student may not understand the meaning of the question.	Student gives an incorrect answer with a reasonable attempt. Student may misread one scale.	Student gives a partially correct answer: • Accurately reads the scales. • Writes the addition equation correctly.	Student correctly answers the question: • Accurately reads scales, almonds = 223 g, raisins = 355 g. • Writes the addition expression 223 + 355. • Solves with 578 g.
3 3.MD.2	Student gives an incorrect answer. Attempt shows the student may not understand the meaning of the question.	Student gives an incorrect answer with a reasonable attempt: • Accurately reads bottle at 2 liters. • Attempts to calculate bottles.	Student gives a partially correct answer: • Accurately reads bottle. • Calculates 9 bottles.	Student correctly answers the question: • Accurately reads bottle. • Draws and labels tape diagram. • Calculates 9 bottles.

Module 2: Place Value and Problem Solving with Units of Measure

A Progression Toward Mastery

4 **3.MD.2** **3.NBT.2**	Student gives an incorrect answer. Attempt shows the student may not understand the meaning of the question.	Student gives an incorrect answer, reasonable attempt: ▪ Accurately reads scale at 744 g. ▪ Attempts to solve.	Student gives a partially correct answer: ▪ Accurately reads scale at 744 g. ▪ Solves with 224 g.	Student correctly answers the question: ▪ Accurately reads scale at 744 g. ▪ Writes a number sentence to calculate the weight of the peas, 224 g. Possible number sentence: 968 – 744 = 224.
5 **3.NBT.2** **3.MD.1** **3.MD.2**	Student gives an incorrect answer. Attempt shows the student may not understand the meaning of the questions.	Student gives an incorrect answer with a reasonable attempt. Student misreads scale but calculates other parts of the problem correctly based on mistake.	Student gives a partially correct answer. All parts are correct besides Part (c), which may not be correctly calculated.	Student correctly answers each part of the question: a. Accurately reads scale at 3 kg. b. Estimates 21 kg. c. Estimates 12 kg. d. Calculates 46 minutes.

A STORY OF UNITS — Mid-Module Assessment Task 3•2

Name __Gina_____ Date _____

1. Fatima runs errands.

   a. The clock to the right shows what time she leaves home. What time does she leave?

   Fatima leaves at 2:07 p.m.

   Fatima leaves home.

   b. It takes Fatima 17 minutes to go from her home to the market. Use the number line below to show what time she gets to the market.

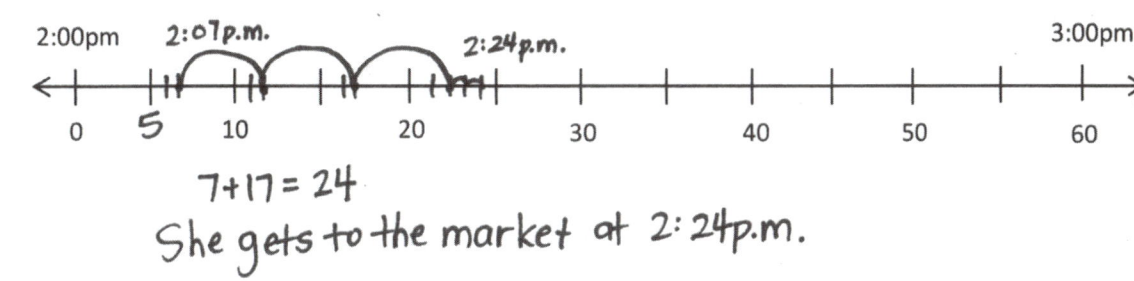

   2:00pm 2:07p.m.   2:24p.m.   3:00pm
   |----|--|--|--|----|----|----|----|----|----|----|----|
   0   5   10   20   30   40   50   60

   7 + 17 = 24
   She gets to the market at 2:24 p.m.

   c. The clock to the right shows what time Fatima leaves the market. What time does she leave the market?

   Fatima leaves the market at 2:53 p.m.

   Fatima leaves the market.

   d. How long does Fatima spend at the market?

   53 − 24 = 50 − 20 − 1
   50 3   20 3 1 = 29

   Fatima is at the store 29 minutes.

2. At the market, Fatima uses a scale to weigh a bag of almonds and a bag of raisins, shown below. What is the total weight of the almonds and raisins?

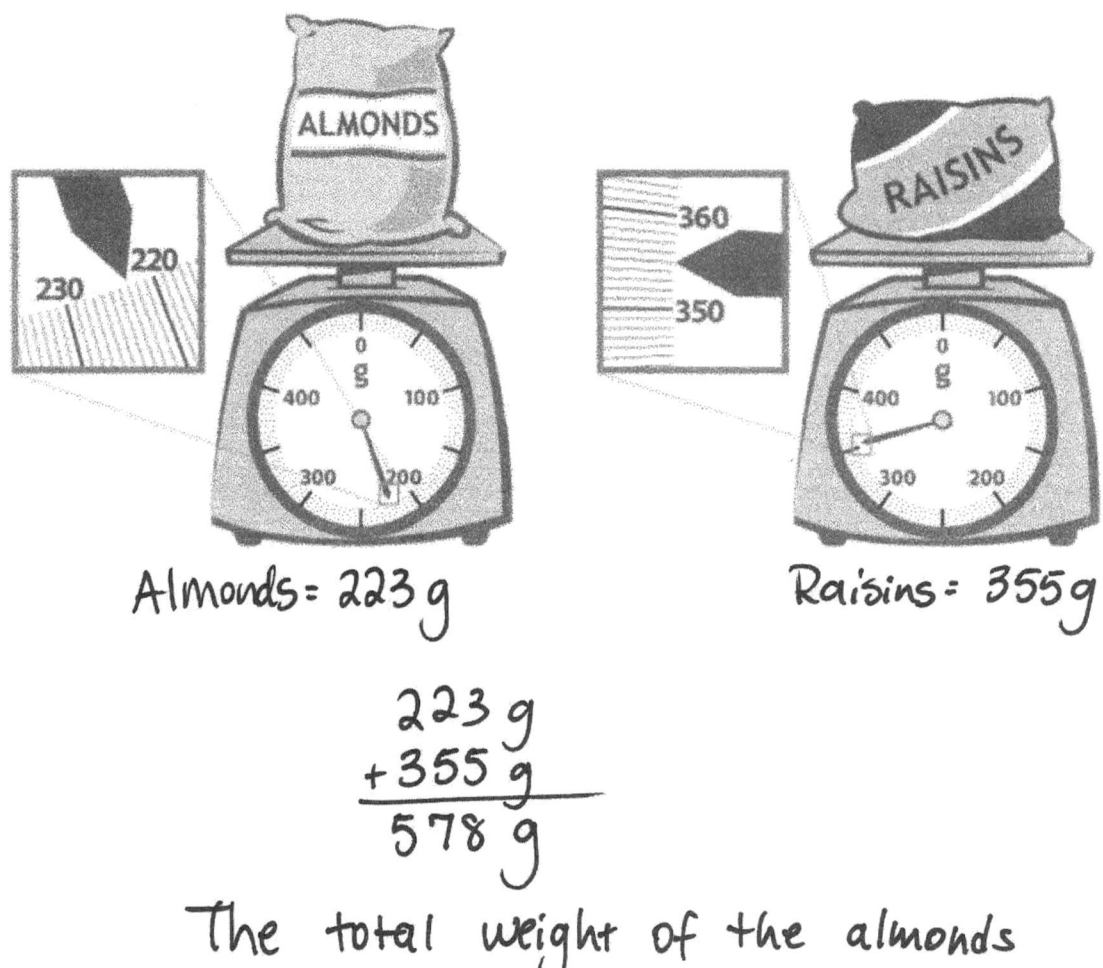

Almonds = 223 g

Raisins = 355 g

```
  223 g
 +355 g
 ─────
  578 g
```

The total weight of the almonds and the raisins is 578 grams.

3. The amount of juice in 1 bottle is shown to the right. Fatima needs 18 liters for a party. Draw and label a tape diagram to find how many bottles of juice she should buy.

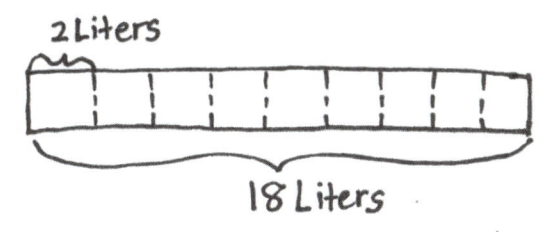

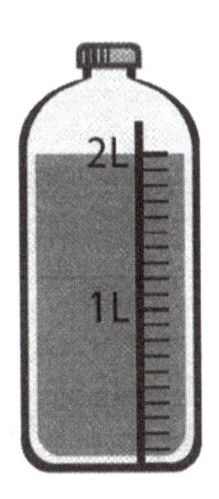

2 Liters

[tape diagram with 9 sections, braced as 18 Liters]

$18 \div 2 = 9$

Fatima needs to buy 9 bottles of juice.

4. Altogether Fatima's lettuce, broccoli and peas weigh 968g. The total weight of her lettuce and broccoli is shown to the right. Write and solve a number sentence to find how much the peas weigh.

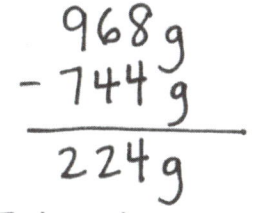

$$\begin{array}{r} 968g \\ -744g \\ \hline 224g \end{array}$$

Fatima's peas weigh 224 grams.

5. Fatima weighs a watermelon, shown to the right.
   a. How much does the watermelon weigh?

   The watermelon weighs 3 kg.

   b. Leaving the store Fatima thinks, "Each bag of groceries seems as heavy as a watermelon!" Use Fatima's idea about the weight of the watermelon to estimate the total weight of 7 bags.

   7 × 3 kg = 21 kg

   She estimates the bags weigh about 21 kg altogether.

   c. The grocer helps carry about 9 kilograms. Fatima carries the rest. Estimate how many kilograms of groceries Fatima carries.

   21 kg − 9 kg = 12 kg   Fatima carries about 12 kg of groceries.
   11 10
   10 − 9 = 1
   11 + 1 = 12

   d. It takes Fatima 12 minutes to drive to the bank after she leaves the store, then 34 more minutes to drive home. How many minutes does Fatima drive after she leaves the store?

   12 minutes + 34 minutes = 46 minutes

   Fatima drives for 46 minutes.

A STORY OF UNITS

Mathematics Curriculum

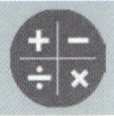

GRADE 3 • MODULE 2

Topic C
Rounding to the Nearest Ten and Hundred

3.NBT.1, 3.MD.1, 3.MD.2

Focus Standards:	3.NBT.1	Use place value understanding to round whole numbers to the nearest 10 or 100.
	3.MD.1	Tell and write time to the nearest minute and measure time intervals in minutes. Solve word problems involving addition and subtraction of time intervals in minutes, e.g., by representing the problem on a number line diagram.
	3.MD.2	Measure and estimate liquid volumes and masses of objects using standard units of grams (g), kilograms (kg), and liters (l). Add, subtract, multiply, or divide to solve one-step word problems involving masses or volumes that are given in the same units, e.g., by using drawings (such as a beaker with a measurement scale) to represent the problem.
Instructional Days:	3	
Coherence -Links from:	G2–M2	Addition and Subtraction of Length Units
-Links to:	G4–M2	Unit Conversions and Problem Solving with Metric Measurement

Topic C builds on students' Grade 2 work with comparing numbers according to the value of digits in the hundreds, tens, and ones places (**2.NBT.4**). Lesson 12 formally introduces rounding two-digit numbers to the nearest ten. Rounding to the leftmost unit usually presents the least challenging type of estimate for students, and so here the sequence begins. Students measure two-digit intervals of minutes and metric measurements, and then use place value understanding to round. They understand that when moving to the right across the places in a number, the digits represent smaller units. Intervals of minutes and metric measurements provide natural contexts for estimation. The number line, presented vertically, provides a new perspective on a familiar tool.

Students continue to use the vertical number line in Lessons 13 and 14. Their confidence with this tool by the end of Topic C lays the foundation for further work in Grades 4 and 5 (**4.NBT.3, 5.NBT.4**). In Lesson 13, the inclusion of rounding three-digit numbers to the nearest ten adds new complexity to the previous day's learning. Lesson 14 concludes the module as students round three- and four-digit numbers to the nearest hundred.

A STORY OF UNITS

Topic C 3•2

A Teaching Sequence Toward Mastery of Rounding to the Nearest Ten and Hundred

Objective 1: Round two-digit measurements to the nearest ten on the vertical number line.
(Lesson 12)

Objective 2: Round two- and three-digit numbers to the nearest ten on the vertical number line.
(Lesson 13)

Objective 3: Round to the nearest hundred on the vertical number line.
(Lesson 14)

Lesson 12

Objective: Round two-digit measurements to the nearest ten on the vertical number line.

Suggested Lesson Structure

- Fluency Practice   (9 minutes)
- Concept Development   (41 minutes)
- Student Debrief   (10 minutes)
- **Total Time**   **(60 minutes)**

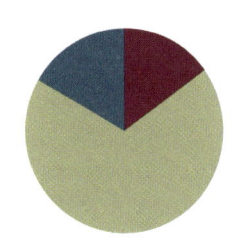

NOTES ON LESSON STRUCTURE:

This lesson does not include an Application Problem but rather uses an extended amount of time for the Problem Set. The Problem Set provides an opportunity for students to apply their newly acquired rounding skills to measurement.

Fluency Practice (9 minutes)

- Rename the Tens 3.NBT.3   (4 minutes)
- Halfway on the Number Line 3.NBT.1   (5 minutes)

Rename the Tens (4 minutes)

Materials: (S) Personal white board

Note: This activity anticipates rounding in Lessons 13 and 14 by reviewing unit form.

T: (Write 9 tens = ____.) Say the number.
S: 90.

Continue with the following possible sequence: 10 tens, 12 tens, 17 tens, 27 tens, 37 tens, 87 tens, 84 tens, and 79 tens.

Halfway on the Number Line (5 minutes)

Materials: (S) Personal white board   *I do*

Note: This activity prepares students to round to the nearest ten in this lesson.

T: (Project a vertical line with endpoints labeled 10 and 20.) What number is halfway between 1 ten and 2 tens?
S: 15.
T: (Write 15, halfway between 10 and 20.)

Repeat process with endpoints labeled 30 and 40.

T: Draw a vertical number line on your personal white board, and make tick marks at each end.

A STORY OF UNITS • Lesson 12 • 3•2

T: (Write 2 tens and 3 tens.) Label the tick marks at each end and at the halfway point.
S: (Label 20 as the bottom point, 30 as the top point, and 25 as the halfway point.)

Continue with 90 and 100.

Concept Development (41 minutes)

Materials: (T) 100 mL beaker, water (S) Personal white board

T: (Show a beaker holding 73 milliliters of water.) This beaker has 73 milliliters of water in it. Show the amount on a vertical number line. Draw a vertical number line, like in today's Fluency Practice. (Model a vertical number line with tick marks for endpoints and a halfway point.)
S: (Draw.)
T: How many tens are in 73?
S: 7 tens!
T: Follow along with me on your board. (To the right of the lowest tick mark, write 70 = 7 tens.)
T: What is 1 more ten than 7 tens?
S: 8 tens!
T: (Write 80 = 8 tens to the right of the top tick mark.)
S: (Label.)
T: What number is halfway between 7 tens and 8 tens?
S: 7 tens and 5 ones, or 75.
T: (Write 75 = 7 tens 5 ones to the right of the halfway point.) Label the halfway point.
S: (Label.)
T: Let's plot 73 on the number line. Remind me, what unit are we plotting on the number line?
S: Milliliters!
T: Say "Stop!" when my finger points to where 73 milliliters should be. (Move finger up the number line from 70 toward 75.)

MP.6

S: Stop!
T: (Plot and label 73 = 7 tens 3 ones.) Now that we know where 73 milliliters is, we can **round** the measurement to the nearest 10 milliliters. Look at your vertical number line. Is 73 milliliters more than halfway or less than halfway between 70 milliliters and 80 milliliters? Tell your partner how you know.

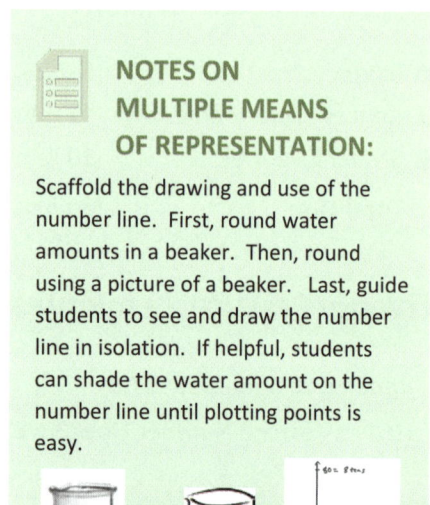

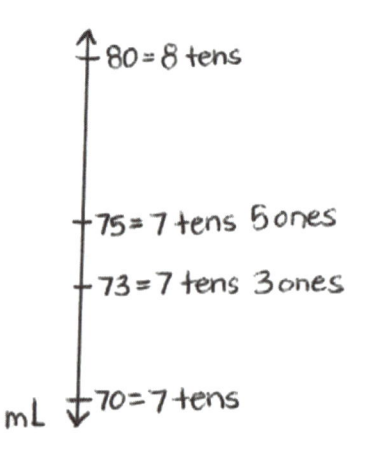

> **NOTES ON MULTIPLE MEANS OF REPRESENTATION:**
>
> Scaffold the drawing and use of the number line. First, round water amounts in a beaker. Then, round using a picture of a beaker. Last, guide students to see and draw the number line in isolation. If helpful, students can shade the water amount on the number line until plotting points is easy.

Lesson 12: Round two-digit measurements to the nearest ten on the vertical number line.

A STORY OF UNITS

Lesson 12 3•2

MP.6

S: 73 milliliters is less than halfway between 70 and 80 milliliters. I know because 3 is less than 5, and 5 marks is halfway. → 73 is 7 away from 80 but only 3 away from 70.

T: 73 milliliters rounded to the nearest ten is 70 milliliters. Another way to say it is that 73 milliliters is **about** 70 milliliters. *About* means that 70 milliliters is not the exact amount.

Continue with the following possible sequence: 61 centimeters, 38 minutes, and 25 grams. For each example, show how the vertical number line can be used even though the units have changed. Be sure to have a discussion about the convention of rounding numbers that end in 5 up to the next ten.

NOTES ON MULTIPLE MEANS OF REPRESENTATION:

For those students who have trouble conceptualizing *halfway*, demonstrate it using students as models. Two students represent the tens. A third student represents the number that is halfway. A fourth student represents the number being rounded. Discuss: Where does the student being rounded belong? When is the student more than halfway? Less than halfway? To which number would they round?

Problem Set (21 minutes)

Materials: (S) Problem Set, 4 bags of rice (pre-measured at four different weights within 100 g), 4 containers of water (pre-measured with four different liquid volumes within 100 mL), ruler, meter stick, blank paper, new pencil, digital scale measuring grams, 100 mL beaker, demonstration clock, classroom wall clock

Description: Students move through different stations to measure using centimeters, grams, milliliters, and minutes as units. Then, they apply learning from the Concept Development to round each measurement to the nearest ten. Students use a ruler, a clock, a beaker, or a drawn vertical number line as tools for rounding to the nearest ten.

NOTES ON MATERIALS:

Adjust the number of measurement materials at each station (ruler, meter stick, digital scale, beaker, demonstration clock) depending both on what is available and on the number of students working at each station at a given time.

Directions: Work with a partner and move through the following stations to complete the Problem Set. Measure, and then round each measurement to the nearest ten.

- Station 1: Measure and round metric length using centimeters. (Provide the four objects listed in Problem 1 of the Problem Set, rulers, and meter sticks.)
- Station 2: Measure and round weight using grams. (Provide four bags of rice labeled at various weights below 100 grams and digital scales that measure in grams.)
- Station 3: Measure and round liquid volume using milliliters. (Provide four containers of various liquid volumes below 100 milliliters and 100-milliliter beakers for measuring.)
- Station 4: (Ongoing, students update the data for this station at Stations 1–3.) Record the exact time you start working at the first station, then the time you finish working at Stations 1, 2, and 3. Then, round each time to the nearest 10 minutes. (Provide demonstration clocks or have students draw vertical number lines to round.)

Lesson 12: Round two-digit measurements to the nearest ten on the vertical number line.

A STORY OF UNITS											Lesson 12 3•2

Prepare students:

- Explain how to complete the problems using the examples provided in the Problem Set.
- Discuss how to perform the measurements at each station.
- Establish which tools students should use for rounding at each station (or differentiate for individual pairs of students).
- Clarify that students should ignore the numbers after the decimal point if scales measure more accurately than to the nearest gram. Students are rounding whole numbers.

Note: Making an immediate connection between the actual measurement and the rounded measurement helps students see the value of rounding. This activity concretizes the relationship between a given number and its relationship to the tens on either side of it. Students also see that when embedded within specific, real, and varied measurement contexts, 73 milliliters and 73 centimeters (rounded or not) have quite different meanings despite appearing nearly synonymous on the number line. Provide students with the language and guidance to engage in discussions that allow these ideas to surface.

Student Debrief (10 minutes)

Lesson Objective: Round two-digit measurements to the nearest ten on the vertical number line.

The Student Debrief is intended to invite reflection and active processing of the total lesson experience.

Invite students to review their work in the Problem Set. They should compare answers with a partner before going over answers as a class. Look for misconceptions that can be addressed in the Debrief. Guide students in a conversation to debrief the Problem Set and process the lesson.

Any combination of the questions below may be used to lead the discussion.

- Discuss new vocabulary from today's lesson: **round** and **about**.

152   Lesson 12:   Round two-digit measurements to the nearest ten on the vertical number line.

EUREKA MATH™

This work is derived from Eureka Math ™ and licensed by Great Minds. ©2015 Great Minds. eureka-math.org

- Why is a vertical number line a good tool to use for rounding?
- How does labeling the halfway point help you to round?
- How did you round numbers that were the same as the halfway point?
- What are some real-world situations where it would be useful to round and estimate?

Exit Ticket (3 minutes)

After the Student Debrief, instruct students to complete the Exit Ticket. A review of their work will help with assessing students' understanding of the concepts that were presented in today's lesson and planning more effectively for future lessons. The questions may be read aloud to the students.

A STORY OF UNITS   Lesson 12 Problem Set 3•2

Name _____ Date _____

1. Work with a partner. Use a ruler or a meter stick to complete the chart below.

Object	Measurement (in cm)	The object measures between (which two tens)...	Length rounded to the nearest 10 cm
Example: My shoe	23 cm	__20__ and __30__ cm	20 cm
Long side of a desk		_____ and _____ cm	
A new pencil		_____ and _____ cm	
Short side of a piece of paper		_____ and _____ cm	
Long side of a piece of paper		_____ and _____ cm	

2. Work with a partner. Use a digital scale to complete the chart below.

Bag	Measurement (in g)	The bag of rice measures between (which two tens)...	Weight rounded to the nearest 10 g
Example: Bag A	8 g	__0__ and __10__ g	10 g
Bag B		_____ and _____ g	
Bag C		_____ and _____ g	
Bag D		_____ and _____ g	
Bag E		_____ and _____ g	

Lesson 12: Round two-digit measurements to the nearest ten on the vertical number line.

3. Work with a partner. Use a beaker to complete the chart below.

Container	Measurement (in mL)	The container measures between (which two tens)...	Liquid volume rounded to the nearest 10 mL
Example: Container A	33 mL	30 and 40 mL	30 mL
Container B		_____ and _____ mL	
Container C		_____ and _____ mL	
Container D		_____ and _____ mL	
Container E		_____ and _____ mL	

4. Work with a partner. Use a clock to complete the chart below.

Activity	Actual time	The activity measures between (which two tens)...	Time rounded to the nearest 10 minutes
Example: Time we started math	10:03	10:00 and 10:10	10:00
Time I started the Problem Set		_____ and _____	
Time I finished Station 1		_____ and _____	
Time I finished Station 2		_____ and _____	
Time I finished Station 3		_____ and _____	

A STORY OF UNITS   Lesson 12 Exit Ticket 3•2

Name _____   Date _____

The weight of a golf ball is shown below.

a. The golf ball weighs _____.

b. Round the weight of the golf ball to the nearest ten grams. Model your thinking on the number line.

c. The golf ball weighs about _____.

d. Explain how you used the halfway point on the number line to round to the nearest ten grams.

Lesson 12: Round two-digit measurements to the nearest ten on the vertical number line.

A STORY OF UNITS

Lesson 12 Homework 3•2

Name _____ Date _____

1. Complete the chart. Choose objects, and use a ruler or meter stick to complete the last two on your own.

Object	Measurement (in cm)	The object measures between (which two tens)...	Length rounded to the nearest 10 cm
Length of desk	66 cm	_____ and _____ cm	
Width of desk	48 cm	_____ and _____ cm	
Width of door	81 cm	_____ and _____ cm	
		_____ and _____ cm	
		_____ and _____ cm	

2. Gym class ends at 10:27 a.m. Round the time to the nearest 10 minutes.

Gym class ends at about _____ a.m.

3. Measure the liquid in the beaker to the nearest 10 milliliters.

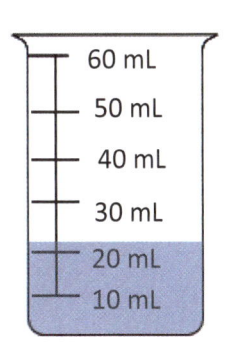

There are about _____ milliliters in the beaker.

Lesson 12: Round two-digit measurements to the nearest ten on the vertical number line.

4. Mrs. Santos' weight is shown on the scale. Round her weight to the nearest 10 kilograms.

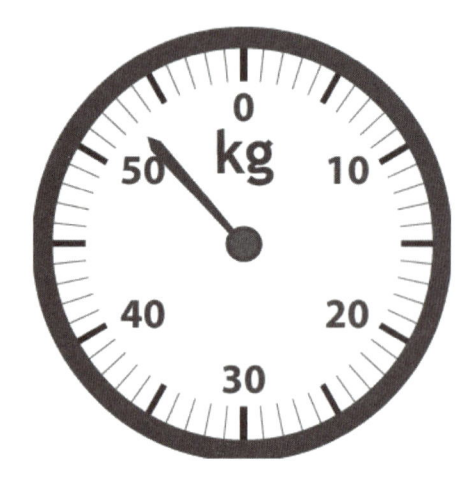

Mrs. Santos' weight is _____ kilograms.

Mrs. Santos weighs about _____ kilograms.

5. A zookeeper weighs a chimp. Round the chimp's weight to the nearest 10 kilograms.

The chimp's weight is _____ kilograms.

The chimp weighs about _____ kilograms.

Lesson 13

Objective: Round two- and three-digit numbers to the nearest ten on the vertical number line.

Suggested Lesson Structure

- Fluency Practice   (13 minutes)
- Application Problem   (7 minutes)
- Concept Development   (30 minutes)
- Student Debrief   (10 minutes)

Total Time   **(60 minutes)**

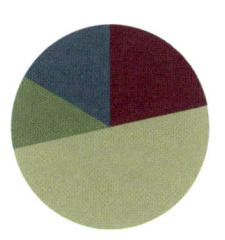

Fluency Practice (13 minutes)

- Group Counting **3.OA.1**   (4 minutes)
- Rename the Tens **3.NBT.3**   (4 minutes)
- Halfway on the Number Line **3.NBT.1** (5 minutes)

Group Counting (4 minutes)

Note: Group counting reviews interpreting multiplication as repeated addition. It reviews foundational strategies for multiplication from Module 1 and anticipates Module 3.

Direct students to count forward and backward, occasionally changing the direction of the count:

- Threes to 30
- Fours to 40
- Sixes to 60
- Sevens to 70
- Eights to 80
- Nines to 90

As students' fluency with skip-counting improves, help them make a connection to multiplication by tracking the number of groups they count using their fingers.

A STORY OF UNITS   Lesson 13 3•2

Rename the Tens (4 minutes)

Note: This activity prepares students for rounding in this lesson and anticipates the work in Lesson 14 where students round numbers to the nearest hundred on the number line.

- T: (Write 9 tens = ____.) Say the number.
- S: 90.

Continue with the following possible sequence: 10 tens, 20 tens, 80 tens, 63 tens, and 52 tens.

Halfway on the Number Line (5 minutes)

Note: This activity reviews rounding using a vertical number line from Lesson 12.

- T: (Project a vertical line with endpoints labeled 30 and 40.) What number is halfway between 3 tens and 4 tens?
- S: 35.
- T: (Write 35 halfway between 30 and 40.)

Continue with the following possible sequence: 130 and 140, 830 and 840, and 560 and 570.

Application Problem (7 minutes)

The school ballet recital begins at 12:17 p.m. and ends at 12:45 p.m. How many minutes long is the ballet recital?

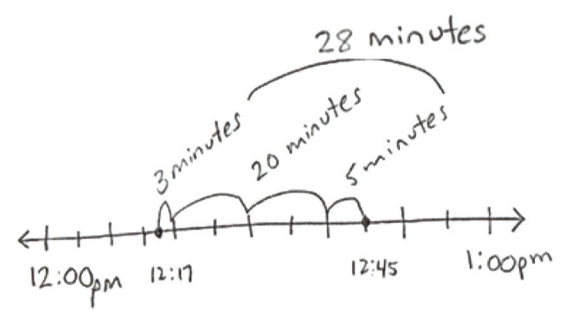

[Number line diagram showing 12:00pm, 12:17, 12:45, 1:00pm with arcs: 3 minutes, 20 minutes, 5 minutes, totaling 28 minutes]

20 + 8 = 28 minutes.

The ballet recital took 28 minutes.

Note: This problem reviews finding intervals of minutes from Topic A and leads directly into rounding intervals of minutes to the nearest ten in this lesson. Encourage students to share and discuss simplifying strategies they may have used to solve. Possible strategies:

- Count by ones from 12:17 to 12:20 and then by fives to 12:45.
- Count by tens and ones, 12:27, 12:37, plus 8 minutes.
- Subtract 17 minutes from 45 minutes.

A STORY OF UNITS

Lesson 13 3•2

Concept Development (30 minutes)

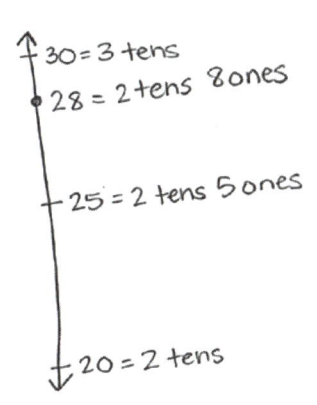

Materials: (T) Place value cards (S) Personal white board

Problem 1: Round two-digit measurements to the nearest ten.

- T: Let's round 28 minutes to the nearest 10 minutes.
- T: How many tens are in 28? (Show place value cards for 28.)
- S: 2 tens! (Pull apart the cards to show the 2 tens as 20. Perhaps cover the zero in the ones to clarify the interpretation of 20 as 2 tens.)
- T: Draw a tick mark near the bottom of the number line. To the right, label it 20 = 2 tens.
- S: (Draw and label 20 = 2 tens.)
- T: What is 1 more ten than 2 tens? 
- S: 3 tens! (Show the place value card for 30 or 3 tens. Again, cover the zero to help clarify.)
- T: Draw a tick mark near the top of the number line. To the right, label it 30 = 3 tens.
- S: (Draw and label 30 = 3 tens.)
- T: What number is halfway between 20 and 30? 
- S: 25.
- T: In unit form, what number is halfway between 2 tens and 3 tens?
- S: 2 tens 5 ones. 
- T: (Show 2 tens 5 ones with the place value cards.) Estimate to draw a tick mark halfway between 20 and 30. Label it 25 = 2 tens 5 ones.
- S: (Draw and label 25 = 2 tens 5 ones.)
- T: When you look at your vertical number line, is 28 more than halfway or less than halfway between 20 and 30? Turn and talk to a partner about how you know. Then plot it on the number line.
- S: 28 is more than halfway between 2 tens and 3 tens. → I know because 28 is more than 25, and 25 is halfway. → I know because 5 ones is halfway, and 8 is more than 5.
- T: What is 28 rounded to the nearest ten?
- S: 30.
- T: Tell me in unit form.
- S: 2 tens 8 ones rounded to the nearest ten is 3 tens.
- T: Let's go back to our Application Problem. How would you round to answer the question, "About how long was the ballet recital?" Discuss with a partner.
- S: The ballet recital took about 30 minutes. → Rounded to the nearest ten, the ballet recital took 30 minutes.

Continue with rounding 17 milliliters to the nearest ten. (Leave the number line used for this on the board. It will be used in Problem 2.)

NOTES ON MULTIPLE MEANS OF ENGAGEMENT:

Alternatively, challenge students who round with automaticity to quickly round 28 minutes to the nearest 10 minutes (without the number line). Students can then write their own word problem for rounding 17 milliliters or 17 minutes.

Lesson 13: Round two- and three-digit numbers to the nearest ten on the vertical number line.

Problem 2: Round three-digit measurements of milliliters to the nearest ten.

T: To round 17 milliliters to the nearest ten, we drew a number line with **endpoints** 1 ten and 2 tens. How will our endpoints change to round 1 *hundred* 17 to the nearest ten? Turn and talk.

S: Each endpoint has to grow by 1 hundred.

T: How many tens are in 1 hundred? (Show the place value card of 100.)

S: 10 tens.

T: When I cover the ones, we see the 10 tens. (Put your hand over the zero in the ones place.)

T: What is 1 more ten than 10 tens?

S: 11 tens.

T: (Show the place value cards for 10 tens and then 11 tens, that is, 100 and 110.)

T: (Show 117 with the place value cards.)

T: How many tens are in 117? Turn and talk about how you know.

S: (Track on fingers.) 10, 20, 30, 40, 50, …, 110. Eleven tens. → 17 has 1 ten, so 117 has 10 tens, plus 1 ten makes 11 tens. → 110 has 11 tens. → 100 has 10 tens and one more ten is 11 tens.

T: What is 1 more ten than 11 tens?

S: 12 tens.

T: What is the value of 12 tens?

S: 120.

T: What will we label our bottom endpoint on the number line when we round 117 to the nearest ten?

S: 110 = 11 tens.

T: The top endpoint?

S: 120 = 12 tens.

T: (Draw and label endpoints on the vertical number line.)

T: How should we label our halfway point?

S: 115 = 11 tens 5 ones.

T: (Show 11 tens 5 ones with the place value cards.) On your number line, mark and label the halfway point.

S: (Mark and label the halfway point.)

T: Is 117 more or less than halfway between 110 and 120? Tell your partner how you know.

S: It's closer to 120. 17 is only 3 away from 20, but 7 away from 10. → It's more than halfway between 110 and 120.

A STORY OF UNITS   Lesson 13 3•2

T: Label 117 on your number line now. (Allow time for students to label 117.) What is 117 rounded to the nearest ten? Use a complete sentence.
S: 117 rounded to the nearest ten is 120.
T: Tell me in unit form with tens and ones.
S: 11 tens 7 ones rounded to the nearest ten is 12 tens.
T: What is 17 rounded to the nearest ten?
S: 20.
T: Again, what is 117 rounded to the nearest ten?
S: 120.
T: Remember from telling time that a number line is continuous. The models we drew to round 17 milliliters and 117 milliliters were the same, even though they showed different portions of the number line; corresponding points are 1 hundred milliliters apart. Discuss the similarities and differences between rounding within those two intervals with your partner.
S: All the numbers went in the same place, we just wrote a 1 in front of them all to show they were 1 hundred more. → We still just paid attention to the number of tens. We thought about if 17 was more or less than halfway between 10 and 20.

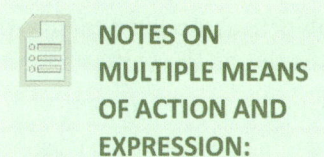

> **NOTES ON MULTIPLE MEANS OF ACTION AND EXPRESSION:**
>
> Reduce the small motor demands of plotting points on a number line by enlarging the number line and offering alternatives to marking with a pencil, such as placing stickers or blocks. Additionally, connect back to yesterday's lesson by using beakers or scales with water or rice.

Continue with rounding the following possible measurements to the nearest ten: 75 mL, 175 mL, 212 g, 315 mL, and 103 kg.

Problem Set (10 minutes)

Students should do their personal best to complete the Problem Set within the allotted 10 minutes. Depending on your class, it may be appropriate to modify the assignment by specifying which problems they work on first. Some problems do not specify a method for solving. Students should solve these problems using the RDW approach used for Application Problems.

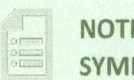

> **NOTES ON SYMBOLS:**
>
> This symbol is used to show that the answer is an approximate: ≈. Before students start work on the Problem Set, call their attention to it and point out the difference between ≈ and =.

Lesson 13: Round two- and three-digit numbers to the nearest ten on the vertical number line.

A STORY OF UNITS

Lesson 13 3•2

Student Debrief (10 minutes)

Lesson Objective: Round two- and three-digit numbers to the nearest ten on the vertical number line.

The Student Debrief is intended to invite reflection and active processing of the total lesson experience.

Invite students to review their solutions for the Problem Set. They should check work by comparing answers with a partner before going over answers as a class. Look for misconceptions or misunderstandings that can be addressed in the Debrief. Guide students in a conversation to debrief the Problem Set and process the lesson.

Any combination of the questions below may be used to lead the discussion.

- What is the same and different about Problems 1(c) and 1(d)? Did you solve the problems differently? Why or why not?
- Look at Problem 1(f). Did the zero in 405 make the problem challenging? Why?
- How did our fluency activities *Rename the Ten* and *Halfway on the Number Line* help with our rounding work today?
- Think back to yesterday's activity where we measured and then rounded at stations. How did that work help you envision the units we worked with today on the number line?

Exit Ticket (3 minutes)

After the Student Debrief, instruct students to complete the Exit Ticket. A review of their work will help with assessing students' understanding of the concepts that were presented in today's lesson and planning more effectively for future lessons. The questions may be read aloud to the students.

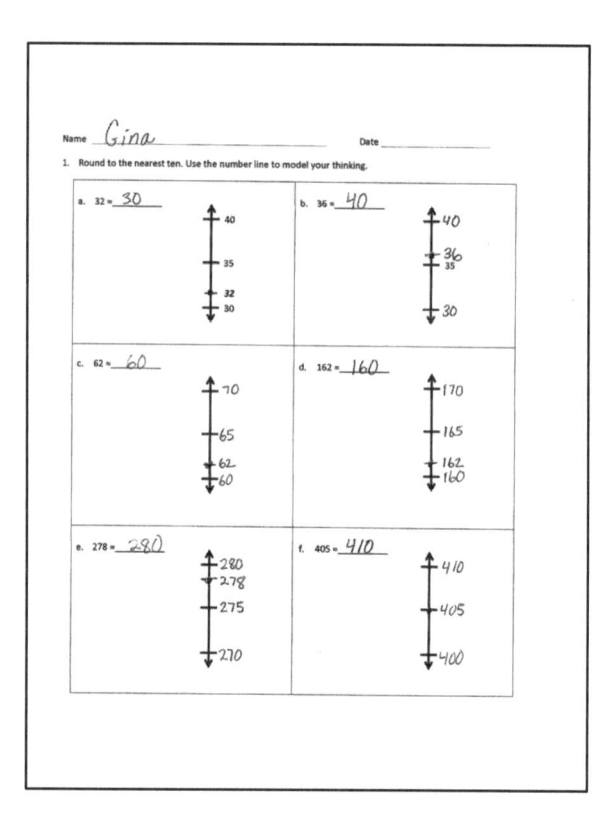

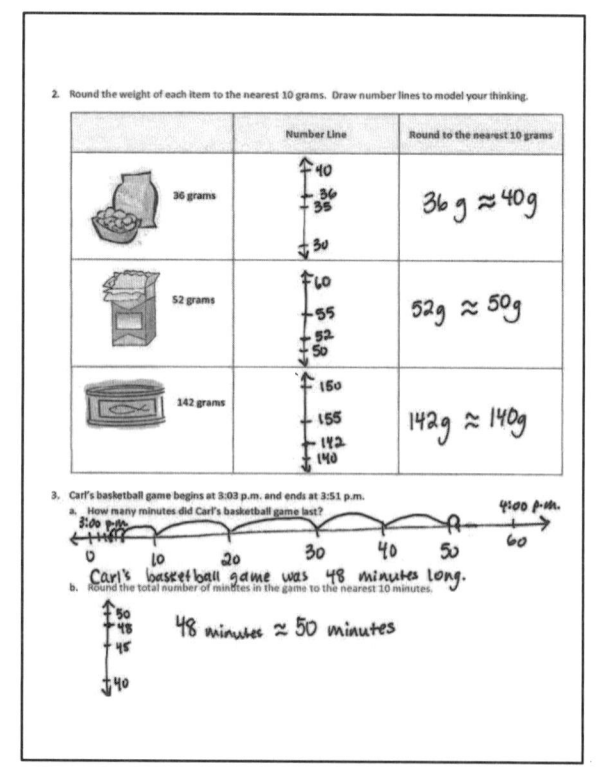

Lesson 13: Round two- and three-digit numbers to the nearest ten on the vertical number line.

Name _____ Date _____

1. Round to the nearest ten. Use the number line to model your thinking.

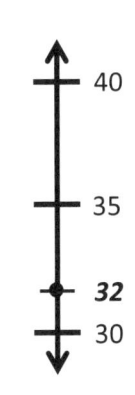

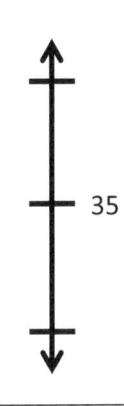

   a. 32 ≈ _____

   — 40
   — 35
   • 32
   — 30

   b. 36 ≈ _____

   — 35

   c. 62 ≈ _____

   d. 162 ≈ _____

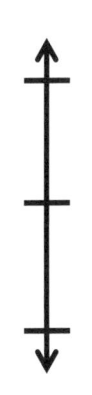

   e. 278 ≈ _____

   f. 405 ≈ _____

2. Round the weight of each item to the nearest 10 grams. Draw number lines to model your thinking.

Item	Number Line	Round to the nearest 10 grams
Chips — 36 grams		
52 grams		
142 grams		

3. Carl's basketball game begins at 3:03 p.m. and ends at 3:51 p.m.

   a. How many minutes did Carl's basketball game last?

   b. Round the total number of minutes in the game to the nearest 10 minutes.

A STORY OF UNITS   Lesson 13 Exit Ticket 3•2

Name _____ Date _____

1. Round to the nearest ten. Use the number line to model your thinking.

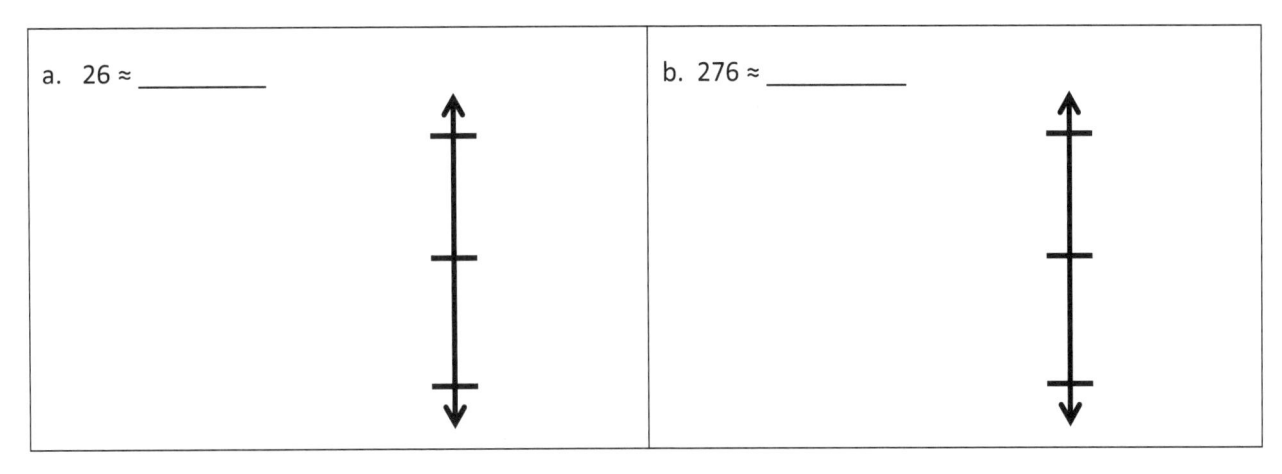

   a. 26 ≈ _____   b. 276 ≈ _____

2. Bobby rounds 603 to the nearest ten. He says it is 610. Is he correct? Why or why not? Use a number line and words to explain your answer.

Name _____   Date _____

1. Round to the nearest ten. Use the number line to model your thinking.

a. 43 ≈ _____

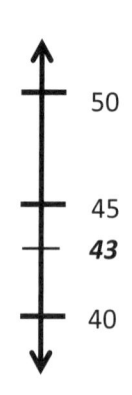

b. 48 ≈ _____

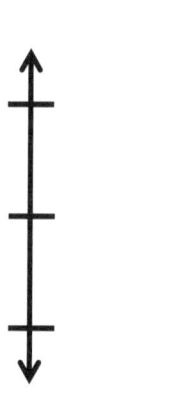

c. 73 ≈ _____

d. 173 ≈ _____

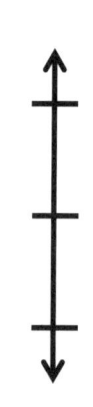

e. 189 ≈ _____

f. 194 ≈ _____

A STORY OF UNITS
Lesson 13 Homework 3•2

2. Round the weight of each item to the nearest 10 grams. Draw number lines to model your thinking.

Item	Number Line	Round to the nearest 10 grams
Cereal bar: 45 grams		
Loaf of bread: 673 grams		

3. The Garden Club plants rows of carrots in the garden. One seed packet weighs 28 grams. Round the total weight of 2 seed packets to the nearest 10 grams. Model your thinking using a number line.

Lesson 13: Round two- and three-digit numbers to the nearest ten on the vertical number line.

Lesson 14

Objective: Round to the nearest hundred on the vertical number line.

Suggested Lesson Structure

- ■ Fluency Practice   (11 minutes)
- ■ Application Problem   (9 minutes)
- ■ Concept Development   (30 minutes)
- ■ Student Debrief   (10 minutes)
- **Total Time**   **(60 minutes)**

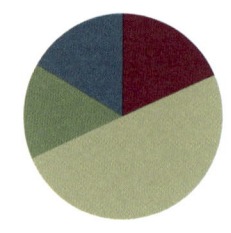

Fluency Practice (11 minutes)

- Sprint: Find the Halfway Point **3.NBT.1**   (9 minutes)
- Rename the Tens **3.NBT.3**   (2 minutes)

Sprint: Find the Halfway Point (9 minutes)

Materials: (S) Find the Halfway Point Sprint

Note: This activity directly supports students' work with rounding by providing practice with finding the halfway point between two numbers.

Rename the Tens (2 minutes)

Note: This activity prepares students for rounding in today's lesson.

- T: (Write 11 tens = _____.) Say the number.
- S: 110.

Continue with the following possible sequence: 19 tens, 20 tens, 28 tens, 30 tens, and 40 tens.

Application Problem (9 minutes)

Materials: (S) Unlabeled place value chart (Template), place value disks (13 hundreds, 10 tens, 8 ones)

Students model the following on the place value chart:

- 10 tens
- 10 hundreds
- 13 tens

A STORY OF UNITS   Lesson 14   3•2

- 13 hundreds
- 13 tens and 8 ones
- 13 hundreds 8 tens 7 ones

Drawn Representation of Place Value Chart and Disks Showing 13 Hundreds 8 Tens 7

MP.6 Note: This problem prepares students for the place value knowledge necessary for Problem 2 in this lesson. They need to understand that there are 13 hundreds in 1387. Through discussion, help students explain the difference between the total number of hundreds in 1387 and the digit in the hundreds place. Use the place value cards to reinforce this discussion if necessary (shown below to the right).

Concept Development (30 minutes)

Materials: (T) Place value cards (S) Personal white board

Problem 1: Round three-digit numbers to the nearest hundred.

T: We've practiced rounding numbers to the nearest ten. Today, let's find 132 grams rounded to the nearest hundred.

T: How many hundreds are in 132 grams? (Show place value cards for 132.)

S: 1 hundred! (Pull apart the cards to show the hundred as 100.)

T: Draw a vertical number line on your personal white board. (Allow students to draw number line.) Draw a tick mark near the bottom of the number line. To the right, label it 100 = 1 hundred.

S: (Draw and label 100 = 1 hundred.)

T: What is 1 more hundred?

S: 2 hundreds! (Show the place value card for 200 or 2 hundreds.)

T: Draw a tick mark near the top of the number line. To the right, label 200 = 2 hundreds.

S: (Draw and label 200 = 2 hundreds.)

T: What number is halfway between 100 and 200?

S: 150.

T: In unit form, what number is halfway between 1 hundred and 2 hundreds?

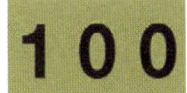

S: 1 hundred 5 tens. (Show with the place value cards.)

T: Estimate to draw a tick mark halfway between 100 and 200. Label it 150 = 1 hundred 5 tens.

S: (Draw and label as 150 = 1 hundred 5 tens.)

T: Estimate to mark and label the location of 132.

S: (Mark and label 132.)

Lesson 14: Round to the nearest hundred on the vertical number line.

T: When you look at your vertical number line, is 132 more than halfway or less than halfway between 100 and 200? Turn and talk to a partner.

S: 132 is less than halfway between 1 hundred and 2 hundreds. → I know because 132 is less than 150, and 150 is halfway. → I know because 5 tens is halfway, and 3 tens is less than 5 tens.

T: 132 grams rounded to the nearest hundred grams is…?

S: 100 grams.

T: Tell me in unit form.

S: 1 hundred 3 tens 2 ones rounded to the nearest hundred is 1 hundred.

✂ Continue with rounding 250 grams and 387 milliliters to the nearest hundred. (Leave the number line for 387 milliliters on the board. It will be used in Problem 2.)

Problem 2: Round four-digit numbers to the nearest hundred.

T: To round 387 milliliters to the nearest hundred, we drew a number line with endpoints 3 hundreds and 4 hundreds. Suppose we round 1,387 milliliters to the nearest hundred. How many hundreds are in 1,387?

S: 13 hundreds.

T: What is 1 more hundred?

S: 14 hundreds.

T: (Draw a vertical number line with endpoints labeled 13 hundreds and 14 hundreds next to the number line for 387.) Draw my number line on your board. Then, work with your partner to estimate, mark, and label the halfway point, as well as the location of 1,387.

S: (Mark and label 13 hundreds 5 tens and 1,387.)

T: Is 1,387 more than halfway or less than halfway between 13 hundreds and 14 hundreds?

S: It's more than halfway.

T: Then, what is 1,387 milliliters rounded to the nearest hundred milliliters?

S: 14 hundred milliliters.

Continue using the following possible sequence: 1,582; 2,146; and 3,245.

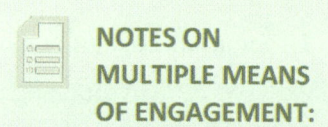

NOTES ON MULTIPLE MEANS OF ENGAGEMENT:

Support students as they locate points on the number line by allowing them to count by tens and mark all points between 1,300 and 1,400.

Alternatively, challenge students to offer three other numbers similar to 2,146 that would be rounded to 2,100.

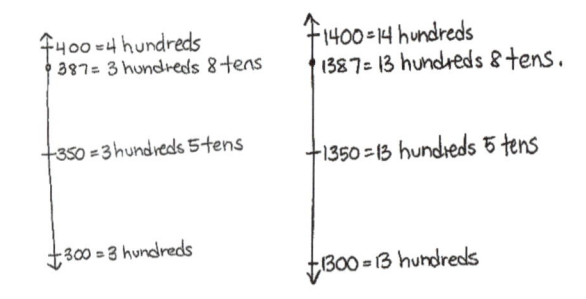

Problem Set (10 minutes)

Students should do their personal best to complete the Problem Set within the allotted 10 minutes. For some classes, it may be appropriate to modify the assignment by specifying which problems they work on first. Some problems do not specify a method for solving. Students should solve these problems using the RDW approach used for Application Problems.

Student Debrief (10 minutes)

Lesson Objective: Round to the nearest hundred on the vertical number line.

The Student Debrief is intended to invite reflection and active processing of the total lesson experience.

Invite students to review their solutions for the Problem Set. They should check work by comparing answers with a partner before going over answers as a class. Look for misconceptions or misunderstandings that can be addressed in the Debrief. Guide students in a conversation to debrief the Problem Set and process the lesson.

Any combination of the questions below may be used to lead the discussion.

- Have students share their explanations for Problem 4, particularly if there is disagreement.
- What strategies did you use to solve Problem 3?
- How is the procedure for rounding to the nearest hundred the same or different for three-digit and four-digit numbers?
- How is rounding to the nearest hundred different from rounding to the nearest ten?

Exit Ticket (3 minutes)

After the Student Debrief, instruct students to complete the Exit Ticket. A review of their work will help with assessing students' understanding of the concepts that were presented in today's lesson and planning more effectively for future lessons. The questions may be read aloud to the students.

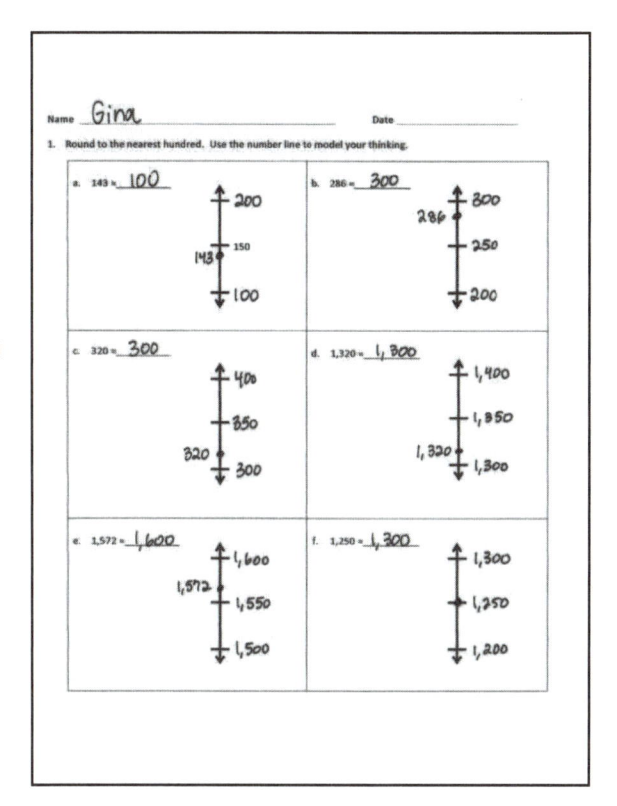

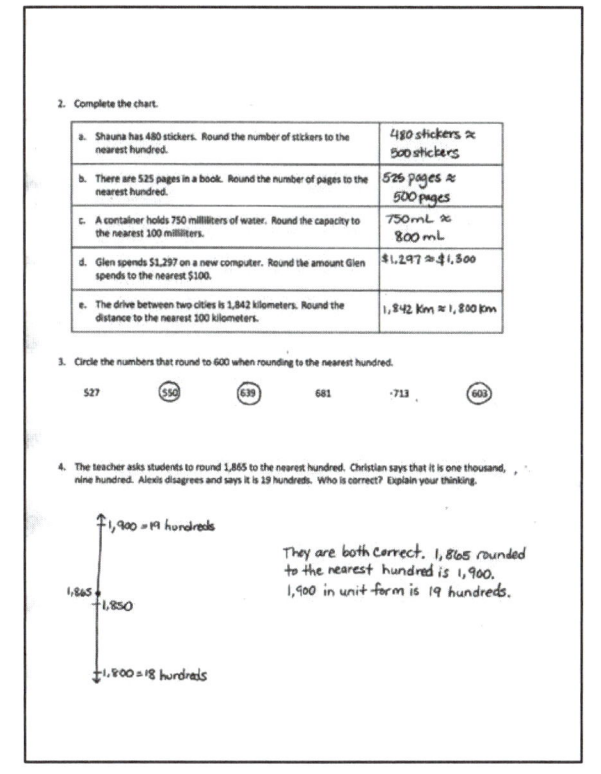

A

Number Correct: _____

Find the Halfway Point

1.	0	_____	10
2.	10	_____	20
3.	20	_____	30
4.	70	_____	80
5.	80	_____	70
6.	40	_____	50
7.	50	_____	40
8.	30	_____	40
9.	40	_____	30
10.	70	_____	60
11.	60	_____	70
12.	80	_____	90
13.	90	_____	100
14.	100	_____	90
15.	90	_____	80
16.	50	_____	60
17.	150	_____	160
18.	250	_____	260
19.	750	_____	760
20.	760	_____	750
21.	80	_____	90
22.	180	_____	190

23.	280	_____	290
24.	580	_____	590
25.	590	_____	580
26.	30	_____	40
27.	930	_____	940
28.	70	_____	60
29.	470	_____	460
30.	90	_____	100
31.	890	_____	900
32.	990	_____	1,000
33.	1,000	_____	1,010
34.	70	_____	80
35.	1,070	_____	1,080
36.	1,570	_____	1,580
37.	480	_____	490
38.	1,480	_____	1,490
39.	1,080	_____	1,090
40.	360	_____	350
41.	1,790	_____	1,780
42.	400	_____	390
43.	1,840	_____	1,830
44.	1,110	_____	1,100

Lesson 14: Round to the nearest hundred on the vertical number line.

A STORY OF UNITS

Lesson 14 Sprint 3•2

Number Correct: _____

Improvement: _____

B

Find the Halfway Point

1.	10	_____	20	23.	270	_____	280
2.	20	_____	30	24.	670	_____	680
3.	30	_____	40	25.	680	_____	670
4.	60	_____	70	26.	20	_____	30
5.	70	_____	60	27.	920	_____	930
6.	50	_____	60	28.	60	_____	50
7.	60	_____	50	29.	460	_____	450
8.	40	_____	50	30.	90	_____	100
9.	50	_____	40	31.	890	_____	900
10.	80	_____	70	32.	990	_____	1,000
11.	70	_____	80	33.	1,000	_____	1,010
12.	80	_____	90	34.	20	_____	30
13.	90	_____	100	35.	1,020	_____	1,030
14.	100	_____	90	36.	1,520	_____	1,530
15.	90	_____	80	37.	380	_____	390
16.	60	_____	70	38.	1,380	_____	1,390
17.	160	_____	170	39.	1,080	_____	1,090
18.	260	_____	270	40.	760	_____	750
19.	560	_____	570	41.	1,690	_____	1,680
20.	570	_____	560	42.	300	_____	290
21.	70	_____	80	43.	1,850	_____	1,840
22.	170	_____	180	44.	1,220	_____	1,210

EUREKA MATH

Lesson 14: Round to the nearest hundred on the vertical number line.

A STORY OF UNITS

Lesson 14 Problem Set 3•2

Name _____   Date _____

1. Round to the nearest hundred. Use the number line to model your thinking.

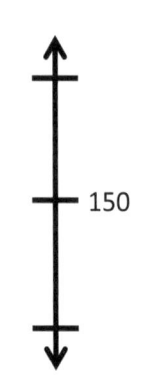

a. 143 ≈ _____

— 150

b. 286 ≈ _____

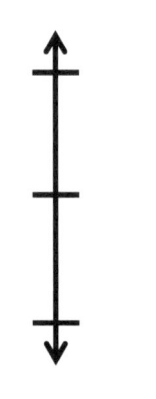

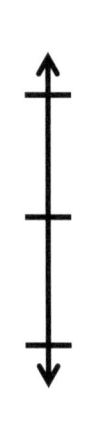

c. 320 ≈ _____

d. 1,320 ≈ _____

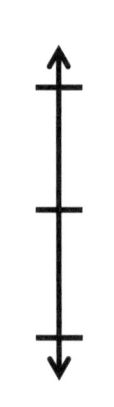

e. 1,572 ≈ _____

f. 1,250 ≈ _____

2. Complete the chart.

a. Shauna has 480 stickers. Round the number of stickers to the nearest hundred.	
b. There are 525 pages in a book. Round the number of pages to the nearest hundred.	
c. A container holds 750 milliliters of water. Round the capacity to the nearest 100 milliliters.	
d. Glen spends $1,297 on a new computer. Round the amount Glen spends to the nearest $100.	
e. The drive between two cities is 1,842 kilometers. Round the distance to the nearest 100 kilometers.	

3. Circle the numbers that round to 600 when rounding to the nearest hundred.

   527   550   639   681   713   603

4. The teacher asks students to round 1,865 to the nearest hundred. Christian says that it is one thousand, nine hundred. Alexis disagrees and says it is 19 hundreds. Who is correct? Explain your thinking.

A STORY OF UNITS

Lesson 14 Exit Ticket 3•2

Name _____ Date _____

1. Round to the nearest hundred. Use the number line to model your thinking.

   a. 137 ≈ _____

   b. 1,761 ≈ _____

2. There are 685 people at the basketball game. Draw a vertical number line to round the number of people to the nearest hundred people.

A STORY OF UNITS   Lesson 14 Homework 3•2

Name _____   Date _____

1. Round to the nearest hundred. Use the number line to model your thinking.

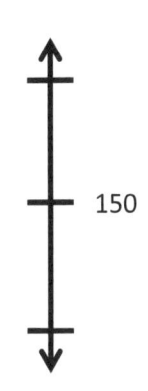

a. 156 ≈ _____

— 150

b. 342 ≈ _____

c. 260 ≈ _____

d. 1,260 ≈ _____

e. 1,685 ≈ _____

f. 1,804 ≈ _____

Lesson 14:   Round to the nearest hundred on the vertical number line.

2. Complete the chart.

a.	Luis has 217 baseball cards. Round the number of cards Luis has to the nearest hundred.	
b.	There were 462 people sitting in the audience. Round the number of people to the nearest hundred.	
c.	A bottle of juice holds 386 milliliters. Round the capacity to the nearest 100 milliliters.	
d.	A book weighs 727 grams. Round the weight to the nearest 100 grams.	
e.	Joanie's parents spent $1,260 on two plane tickets. Round the total to the nearest $100.	

3. Circle the numbers that round to 400 when rounding to the nearest hundred.

   368   342   420   492   449   464

4. There are 1,525 pages in a book. Julia and Kim round the number of pages to the nearest hundred. Julia says it is one thousand, five hundred. Kim says it is 15 hundreds. Who is correct? Explain your thinking.

unlabeled place value chart

A STORY OF UNITS

Mathematics Curriculum

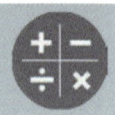

GRADE 3

GRADE 3 • MODULE 2

Topic D
Two- and Three-Digit Measurement Addition Using the Standard Algorithm

3.NBT.2, 3.NBT.1, 3.MD.1, 3.MD.2

Focus Standard:	3.NBT.2	Fluently add and subtract within 1000 using strategies and algorithms based on place value, properties of operations, and/or the relationship between addition and subtraction.
Instructional Days:	3	
Coherence -Links from:	G2–M2	Addition and Subtraction of Length Units
	G2–M5	Addition and Subtraction Within 1000 with Word Problems to 100
-Links to:	G4–M1	Place Value, Rounding, and Algorithms for Addition and Subtraction

In Topic D, students revisit the standard algorithm for addition, which was first introduced in Grade 2 (**2.NBT.7**). In this topic, they add two- and three-digit metric measurements and intervals of minutes within 1 hour. Lesson 15 guides students to apply the place value concepts they practiced with rounding to model composing larger units once on the place value chart. They use the words *bundle* and *rename* as they add like base ten units, working across the numbers unit by unit (ones with ones, tens with tens, hundreds with hundreds). As the lesson progresses, students transition away from modeling on the place value chart and move toward using the standard algorithm.

Lesson 16 adds complexity to the previous day's learning by presenting problems that require students to compose larger units twice. Again, students begin by modeling on the place value chart, this time renaming both the ones and tens places. Lesson 17 culminates the topic with applying addition involving renaming to solving measurement word problems. Students draw tape diagrams to model problems. They round to estimate the sums of measurements and then solve problems using the standard algorithm. By comparing their estimates with precise calculations, students assess the reasonableness of their solutions.

A STORY OF UNITS

Topic D 3•2

A Teaching Sequence Toward Mastery of Two- and Three-Digit Measurement Addition Using the Standard Algorithm

Objective 1: Add measurements using the standard algorithm to compose larger units once.
(Lesson 15)

Objective 2: Add measurements using the standard algorithm to compose larger units twice.
(Lesson 16)

Objective 3: Estimate sums by rounding and apply to solve measurement word problems.
(Lesson 17)

Lesson 15

Objective: Add measurements using the standard algorithm to compose larger units once.

Suggested Lesson Structure

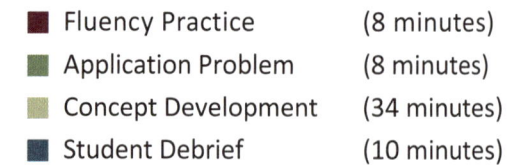

- Fluency Practice   (8 minutes)
- Application Problem   (8 minutes)
- Concept Development   (34 minutes)
- Student Debrief   (10 minutes)
- **Total Time**   **(60 minutes)**

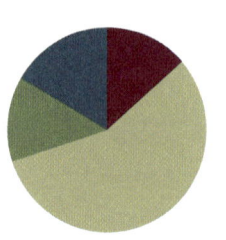

Fluency Practice (8 minutes)

- Part–Whole with Measurement Units **3.MD.2**   (3 minutes)
- Round Three- and Four-Digit Numbers **3.NBT.1**   (5 minutes)

Part–Whole with Measurement Units (3 minutes)

Materials: (S) Personal white board

Note: This activity reviews part–whole thinking using measurement units.

- T: There are 100 centimeters in 1 meter. How many centimeters are in 2 meters?
- S: 200 centimeters.
- T: 3 meters?
- S: 300 centimeters.
- T: 8 meters?
- S: 800 centimeters.
- T: (Write 50 minutes + ____ minutes = 1 hour.) There are 60 minutes in 1 hour. On your personal white board, fill in the equation.
- S: (Write 50 minutes + 10 minutes = 1 hour.)

Continue with the following suggested sequence: 30 minutes and 45 minutes.

- T: (Write 800 mL + ____ mL = 1 L.) There are 1,000 milliliters in 1 liter. On your board, fill in the equation.
- S: (Write 800 mL + 200 mL = 1 L.)

Continue with the following suggested sequence: 500 mL, 700 mL, and 250 mL.

T: (Write 1 kg – 500 g = _____ g.) There are 1,000 grams in 1 kilogram. On your board, fill in the equation.
S: (Write 1 kg – 500 g = 500 g.)

Continue with the following suggested sequence: Subtract 300 g, 700 g, and 650 g from 1 kg.

Round Three- and Four-Digit Numbers (5 minutes)

Materials: (S) Personal white board

Note: This activity reviews rounding from Lessons 13 and 14.

T: (Write 87 ≈ ___.) What is 87 rounded to the nearest ten?
S: 90.

Continue with the following possible sequence: 387, 43, 643, 35, and 865.

T: (Write 237 ≈ ___.) 237 is between which 2 hundreds?
S: 200 and 300.
T: On your board, draw a vertical number line. Mark 200 and 300 as your endpoints and label the halfway point.
S: (Label 200 and 300 as endpoints and 250 as the halfway point.)
T: Show where 237 falls on the number line, and then round to the nearest hundred.
S: (Plot 237 between 200 and 250 and write 237 ≈ 200.)

Continue with the following suggested sequence: 1,237; 678; 1,678; 850; 1,850; and 2,361.

Application Problem (8 minutes)

Use mental math to solve these problems. Record your strategy for solving each problem.

a. 46 mL + 5 mL   b. 39 cm + 8 cm   c. 125 g + 7 g   d. 108 L + 4 L

Possible strategies:
  a. 46 mL + 4 mL + 1 mL = 50 mL + 1 mL = 51 mL
  b. 39 cm + 1 cm + 7 cm = 40 cm + 7 cm = 47 cm
  c. 125 g + 5 g + 2 g = 130 g + 2 g = 132 g
  d. 108 L + 2 L + 2 L = 110 L + 2 L = 112 L

Note: This problem is designed to show that mental math can be an efficient strategy even when renaming is required. It also sets up the conversation in the Student Debrief about when and why the standard algorithm is used. Be sure to give students an opportunity to discuss and show how they solved these problems.

A STORY OF UNITS

Lesson 15 3•2

Concept Development (34 minutes)

Materials: (T) 2 beakers, water (S) Unlabeled place value chart (Lesson 14 Template), place value disks, personal white board

Students start with the unlabeled place value chart template in their personal white boards.

Lesson 14 Template

- T: (Show Beaker A with 56 milliliters of water and Beaker B with 27 milliliters of water.) Beaker A has 56 milliliters of water, and Beaker B has 27 milliliters of water. Let's use place value charts and place value disks to find the total milliliters of water in both beakers.
- T: Use place value disks to represent the amount of water from Beaker A on your chart. (Allow time for students to work.)
- T: Record 56 milliliters in the workspace on your personal white board below the place value chart.
- T: Leave the disks for 56 on your chart. Use more disks to represent the amount of water from Beaker B. Place them below your model of 56. (Allow time for students to work.)
- T: In the workspace on your board, use an addition sign to show that you added 27 milliliters to 56 milliliters.
- T: (Point to the place value disks in the ones column.) Six ones plus 7 ones equals …?
- S: 13 ones.
- T: We can change 10 ones for 1 ten. Take 10 ones disks and change them for 1 tens disk. Where do we put the tens disk on the place value chart?
- S: In the tens column.
- T: How many ones do we have now?
- S: 3 ones!
- T: Let's show that same work in the problem we wrote in our workspace on our boards. If you wrote your problem horizontally, rewrite it vertically so that it looks like mine.
- T: (Point to the ones.) 6 ones plus 7 ones equals …?
- S: 13 ones.
- T: Let's rename some ones as tens. How many tens and ones in 13 ones?
- S: 1 ten and 3 ones.

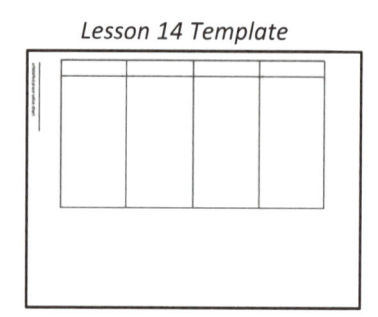

> **NOTES ON MULTIPLE MEANS OF ACTION AND EXPRESSION:**
>
> English language learners and others will benefit from the real-world context, the varied methods for response (personal white boards, models, numbers, etc.), and the introduction to academic math language (*standard algorithm*) at the end of the lesson.

$$\begin{array}{r} 56 \text{ mL} \\ + 27 \text{ mL} \\ \hline \end{array}$$

Lesson 15: Add measurements using the standard algorithm to compose larger units once.

T: This is how we show renaming using the **standard algorithm**. (Write the 1 so that it crosses the line under the tens in the tens place and the 3 below the line in the ones column. This way you write 13 rather than 3 and 1 as separate numbers. Refer to the vertical addition shown to the right.) Show this work on your board.

T: Talk to a partner. How is this work similar to the work we did with the place value disks?

S: (Discuss.)

T: That's right. Renaming in the algorithm is the same as changing with our place value disks.

T: (Point to the place value disks in the tens column.) 5 tens plus 2 tens plus 1 ten equals…?

S: 8 tens!

T: 8 tens 3 ones makes how many milliliters of water in the bowl?

S: 83 milliliters.

T: Let's show that in our problem. (Point to the tens.) 5 tens plus 2 tens plus 1 ten equals…?

S: 8 tens.

T: Record 8 tens below the line in the tens column.

T: What unit do we need to include in our answer?

S: Milliliters!

T: Read the problem with me. (Point and read.) 56 milliliters plus 27 milliliters equals 83 milliliters. We just used the standard algorithm as a tool for solving this problem.

T: How can I check our work using the beaker?

S: Pour the water from one beaker into the other beaker and read the measurement.

T: (Pour.) The amount of water in the beaker is 83 milliliters!

Continue with the following suggested problems:

- *Add to with start unknown:* Lisa draws a line on the board. Marcus shortens the length of the line by erasing 32 centimeters. The total length of the line is now 187 centimeters. How long is the line that Lisa drew?
- *Compare with bigger unknown (start unknown):* John reads for 74 minutes on Wednesday. On Thursday, he reads for 17 more minutes than he read on Wednesday. How many total minutes does John read on Wednesday and Thursday?

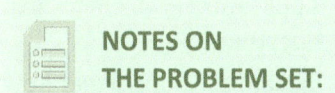

Problem Set (10 minutes)

Students should do their personal best to complete the Problem Set within the allotted 10 minutes. For some classes, it may be appropriate to modify the assignment by specifying which problems they work on first. Some problems do not specify a method for solving. Students should solve these problems using the RDW approach used for Application Problems.

> **NOTES ON THE PROBLEM SET:**
>
> The problems in the Problem Set are written horizontally so that students do not assume that they need to use the standard algorithm to solve. Mental math may be a more efficient strategy in some cases. Invite students to use the algorithm as a strategic tool, purposefully choosing it rather than defaulting to it.

Lesson 15: Add measurements using the standard algorithm to compose larger units once.

Student Debrief (10 minutes)

Lesson Objective: Add measurements using the standard algorithm to compose larger units once.

The Student Debrief is intended to invite reflection and active processing of the total lesson experience.

Invite students to review their solutions for the Problem Set. They should check work by comparing answers with a partner before going over answers as a class. Look for misconceptions or misunderstandings that can be addressed in the Debrief. Guide students in a conversation to debrief the Problem Set and process the lesson.

Any combination of the questions below may be used to lead the discussion.

- Notice the units in Problems 1(j) and 1(k). Both problems use both kilograms and grams. Did having two units in the problem change anything about the way you solved?
- **MP.7** What pattern did you notice between Problems 1(a), 1(b), and 1(c)? How did this pattern help you solve the problems?
- Did you rewrite any of the horizontal problems vertically? Why?
- Which problems did you solve using mental math? The **standard algorithm**? Why did you use the standard algorithm for some problems and mental math for other problems? Think about the strategies you used to solve today's Application Problem to help you answer this question.
- Explain to your partner how you used the standard algorithm to solve Problem 3. Did you rename the ones? Tens? Hundreds?
- Explain to your partner what your tape diagram looked like for Problem 4.
- How are Problems 2 and 4 similar? How are they different from the other problems?

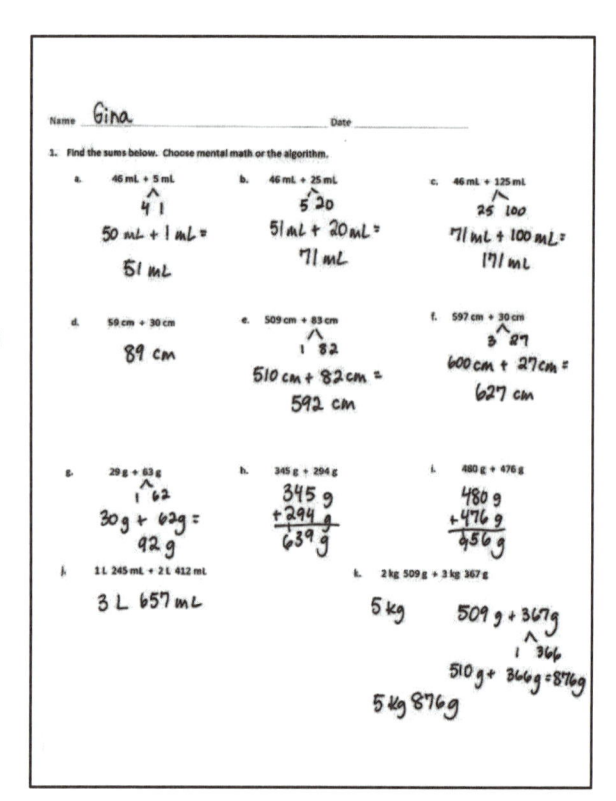

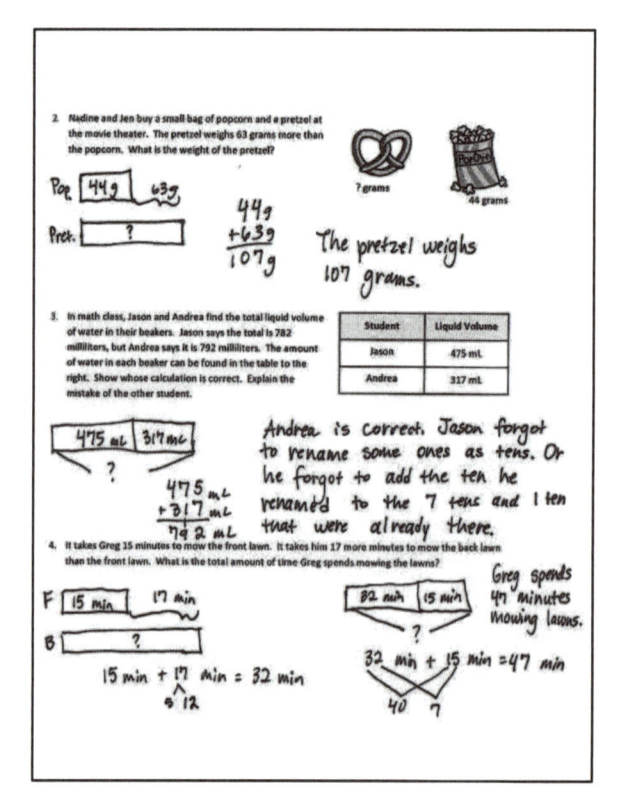

Exit Ticket (3 minutes)

After the Student Debrief, instruct students to complete the Exit Ticket. A review of their work will help with assessing students' understanding of the concepts that were presented in today's lesson and planning more effectively for future lessons. The questions may be read aloud to the students.

Name _____   Date _____

1. Find the sums below. Choose mental math or the algorithm.

   a. 46 mL + 5 mL

   b. 46 mL + 25 mL

   c. 46 mL + 125 mL

   d. 59 cm + 30 cm

   e. 509 cm + 83 cm

   f. 597 cm + 30 cm

   g. 29 g + 63 g

   h. 345 g + 294 g

   i. 480 g + 476 g

   j. 1 L 245 mL + 2 L 412 mL

   k. 2 kg 509 g + 3 kg 367 g

2. Nadine and Jen buy a small bag of popcorn and a pretzel at the movie theater. The pretzel weighs 63 grams more than the popcorn. What is the weight of the pretzel?

? grams

44 grams

3. In math class, Jason and Andrea find the total liquid volume of water in their beakers. Jason says the total is 782 milliliters, but Andrea says it is 792 milliliters. The amount of water in each beaker can be found in the table to the right. Show whose calculation is correct. Explain the mistake of the other student.

Student	Liquid Volume
Jason	475 mL
Andrea	317 mL

4. It takes Greg 15 minutes to mow the front lawn. It takes him 17 more minutes to mow the back lawn than the front lawn. What is the total amount of time Greg spends mowing the lawns?

Name _____ Date _____

1. Find the sums below. Choose mental math or the algorithm.

   a. 24 cm + 36 cm

   b. 562 m + 180 m

   c. 345 km + 239 km

2. Brianna jogs 15 minutes more on Sunday than Saturday. She jogged 26 minutes on Saturday.

   a. How many minutes does she jog on Sunday?

   b. How many minutes does she jog in total?

A STORY OF UNITS   **Lesson 15 Homework 3•2**

Name _____   Date _____

1. Find the sums below. Choose mental math or the algorithm.

   a. 75 cm + 7 cm   c. 362 mL + 229 mL   e. 451 mL + 339 mL

   b. 39 kg + 56 kg   d. 283 g + 92 g   f. 149 L + 331 L

2. The liquid volume of five drinks is shown below.

Drink	Liquid Volume
Apple juice	125 mL
Milk	236 mL
Water	248 mL
Orange juice	174 mL
Fruit punch	208 mL

   a. Jen drinks the apple juice and the water. How many milliliters does she drink in all?

   Jen drinks _____ mL.

   b. Kevin drinks the milk and the fruit punch. How many milliliters does he drink in all?

Lesson 15: Add measurements using the standard algorithm to compose larger units once.

3. There are 75 students in Grade 3. There are 44 more students in Grade 4 than in Grade 3. How many students are in Grade 4?

4. Mr. Green's sunflower grew 29 centimeters in one week. The next week it grew 5 centimeters more than the previous week. What is the total number of centimeters the sunflower grew in 2 weeks?

5. Kylie records the weights of 3 objects as shown below. Which 2 objects can she put on a pan balance to equal the weight of a 460 gram bag? Show how you know.

Paperback Book	Banana	Bar of Soap
343 grams	108 grams	117 grams

Lesson 16

Objective: Add measurements using the standard algorithm to compose larger units twice.

Suggested Lesson Structure

- ■ Fluency Practice   (12 minutes)
- ■ Application Problem   (5 minutes)
- ■ Concept Development   (33 minutes)
- ■ Student Debrief   (10 minutes)
  Total Time   **(60 minutes)**

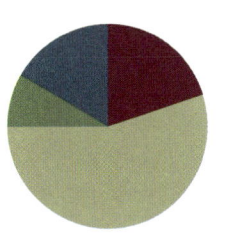

Fluency Practice (12 minutes)

- Part–Whole with Measurement Units **3.MD.2**   (3 minutes)
- Round Three- and Four-Digit Numbers **3.NBT.1** (5 minutes)
- Group Counting **3.OA.1**   (4 minutes)

Part–Whole with Measurement Units (3 minutes)

Materials: (S) Personal white board

Note: This activity reviews part–whole thinking using measurement units.

- T: There are 100 centimeters in 1 meter. How many centimeters are in 4 meters?
- S: 400 centimeters.
- T: 5 meters?
- S: 500 centimeters.
- T: 7 meters?
- S: 700 centimeters.
- T: (Write 30 minutes + ____ minutes = 1 hour.) There are 60 minutes in 1 hour. On your personal white board, fill in the equation.
- S: (Write 30 minutes + 30 minutes = 1 hour.)

Continue with the following suggested sequence: 40 minutes and 25 minutes.

- T: (Write 300 mL + ____ mL = 1 L.) There are 1,000 milliliters in 1 liter. On your board, fill in the equation.
- S: (Write 300 mL + 700 mL = 1 liter.)

Continue with the following suggested sequence: 200 mL, 600 mL, and 550 mL.

Round Three- and Four-Digit Numbers (5 minutes)

Materials: (S) Personal white board

Note: This activity reviews rounding from Lessons 13 and 14.

T: (Write 73 ≈ ___.) What is 73 rounded to the nearest ten?
S: 70.

Repeat the process, varying the numbers.

Group Counting (4 minutes)

Note: Group counting reviews interpreting multiplication as repeated addition. It reviews foundational strategies for multiplication from Module 1 and anticipates Module 3.

Direct students to count forward and backward, occasionally changing the direction of the count:

- Threes to 30
- Fours to 40
- Sixes to 60
- Sevens to 70
- Eights to 80
- Nines to 90

As students' fluency with skip-counting improves, help them make a connection to multiplication by tracking the number of groups they count using their fingers.

Application Problem (5 minutes)

Josh's apple weighs 93 grams. His pear weighs 152 grams. What is the total weight of the apple and the pear?

$$152 \text{ g} + 93 \text{ g} = 245 \text{ g}$$

The total weight of the apple and the pear is 245 grams.

Note: This problem reviews the use of the standard algorithm to compose larger units once.

Lesson 16

Concept Development (33 minutes)

Materials: (T) Bag A of beans (266 grams), Bag B of beans (158 grams), scale that weighs in grams
(S) Personal white board, unlabeled place value chart (Lesson 14 Template), place value disks

Problem 1: Use place value charts, disks, and the standard algorithm to add measurements, composing larger units twice.

Students start with the unlabeled place value chart template in their personal white boards.

Lesson 14 Template

- T: (Show Bags A and B.) Bag A has 266 grams of beans, and Bag B has 158 grams of beans. Let's use our place value charts and place value disks to figure out how many grams of beans we have altogether.
- T: Use disks to represent the weight of the beans in Bag B.
- S: (Put 8 ones disks in the ones column, 5 tens disks in the tens column, and 1 hundreds disk in the hundreds column.)
- T: Record 158 grams in the workspace on your board below the place value chart.
- T: Leave the disks on your chart. Use more disks to represent the weight of the beans in Bag A. Place them below your model of 158.
- S: (Place 6 ones disks, 6 tens disks, and 2 hundreds disks in respective columns.)
- T: In the workspace on your board, use an addition sign to show that you added 266 grams to 158 grams.
- T: (Point to the place value disks in the ones column.) 8 ones plus 6 ones equals…?
- S: 14 ones.
- T: We can change 10 ones for 1 ten. Take 10 ones disks and change them for 1 tens disk. Where do we put the tens disk on the place value chart?
- S: In the tens column.
- T: How many ones do we have now?
- S: 4 ones!
- T: Let's use the standard algorithm to show our work on the place value chart. Use the problem you wrote in the workspace on your board. (Write the problem vertically, as shown.) Be sure your problem is written vertically, like mine.
- T: (Point to the ones in the problem.) 8 ones plus 6 ones equals?
- S: 14 ones.

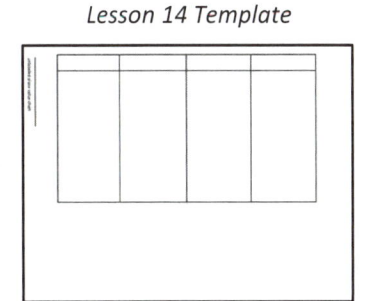

> **NOTES ON MULTIPLE MEANS OF ENGAGEMENT:**
>
> Students working above grade level may be eager to find the sum quickly without using place value disks. Keep these learners engaged by optimizing their choice and autonomy. Request from them an alternative model, such as a tape diagram. They may enjoy offering two more examples of their own in which they use the standard algorithm to compose larger units twice.

$$\begin{array}{r} 158g \\ +266g \\ \hline \end{array}$$

Lesson 16: Add measurements using the standard algorithm to compose larger units twice.

T: Let's rename some ones as tens. How many tens and ones in 14?
S: 1 ten and 4 ones.
T: Just like we practiced yesterday, show that on your problem.
S: (Write the 1 so that it crosses the line under the tens in the tens place and the 4 below the line in the ones column.)
T: (Point to the place value disks in the tens column.) Five tens plus 6 tens plus 1 ten equals?
S: 12 tens!
T: We can change 10 tens for 1 hundred. Take 10 tens disks and change them for 1 hundreds disk. Where do we put the hundreds disk on the place value chart?
S: In the hundreds column.
T: How many tens do we have now?
S: 2 tens!
T: Let's show that in our problem. (Point to the tens in the problem.) Five tens plus 6 tens plus 1 ten equals…?
S: 12 tens.
T: Let's rename some tens as hundreds. How many hundreds and tens in 12 tens?
S: 1 hundred and 2 tens.
T: We show our new hundred just like we showed our new ten before, but this time we put it in the hundreds column because it's a hundred, not a ten. (Write the 1 so that it crosses the line under the hundreds in the hundreds place and the 2 below the line in the tens column.)
T: (Point to the place value disks in the hundreds column.) One hundred plus 2 hundreds plus 1 hundred equals…?
S: 4 hundreds!
T: Four hundreds 2 tens 4 ones makes how many total grams of beans in Bag A and Bag B?
S: 424 grams.
T: Let's show that in our problem. (Point to the hundreds in the problem.) One hundred plus 2 hundreds plus 1 hundred equals…?
S: 4 hundreds!
T: Record 4 hundreds in the hundreds column below the line.
T: What unit do we need to include in our answer?
S: Grams!
T: Read the problem with me. (Point and read.) 158 grams plus 266 grams equals 424 grams.
T: How can I check our work using a scale?
S: Put Bag A and Bag B on the scale and read the measurement.
T: (Put Bags A and B on the scale.) The total weight of the beans is 424 grams!

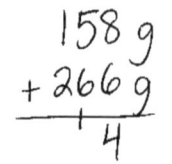

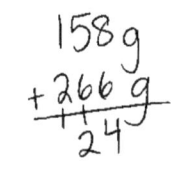

Continue with the following suggested problems:

- *Add to with start unknown:* Jamal had a piece of rope. His brother cut off 47 centimeters and took it! Now, Jamal only has 68 centimeters left. How long was Jamal's rope before his brother cut it?
- *Compare with bigger unknown:* The goldfish aquarium at Sal's Pet Store has 189 liters of water. The guppy aquarium has 94 more liters of water than the goldfish aquarium. How many liters of water are in both aquariums?

Problem 2: Use the partner–coach strategy and the standard algorithm to add measurements, composing larger units twice.

Materials: (S) Problem Set

Students work with a partner and use the partner–coach strategy to complete page 1 of the Problem Set.

Prepare students:

- Explain how to use the partner–coach strategy. (One partner coaches, verbalizing the steps needed to solve the problem, while the other partner writes the solution. Then partners switch roles.)
- Generate a class list of important words that should be included in the coaching conversations (e.g., *ones, tens, hundreds, change, standard algorithm, mental math, rename*). Keep this list posted for students to refer to as they coach each other.

MP.6

Circulate as students work, addressing misconceptions or incorrect work.

Problem Set (5 minutes)

Students should do their personal best to complete page 2 of the Problem Set within the allotted 5 minutes. For some classes, it may be appropriate to modify the assignment by specifying which problems they work on first. Some problems do not specify a method for solving. Students should solve these problems using the RDW approach used for Application Problems.

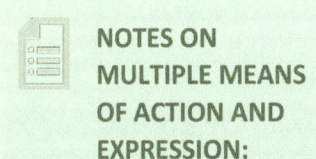

> **NOTES ON MULTIPLE MEANS OF ACTION AND EXPRESSION:**
>
> Help students prepare for successful participation in the Student Debrief. Some may need guidance and support to discover the patterns of Problem 1. Encourage students to read aloud the number sentences in each row and to search for the numbers that repeat.

A STORY OF UNITS

Lesson 16 3•2

Student Debrief (10 minutes)

Lesson Objective: Add measurements using the standard algorithm to compose larger units twice.

The Student Debrief is intended to invite reflection and active processing of the total lesson experience.

Invite students to review their solutions for the Problem Set. They should check work by comparing answers with a partner before going over answers as a class. Look for misconceptions or misunderstandings that can be addressed in the Debrief. Guide students in a conversation to debrief the Problem Set and process the lesson.

Any combination of the questions below may be used to lead the discussion.

- What pattern did you notice between Problems 1(a), 1(b), and 1(c)? How did the pattern help you solve these problems?
- Did you or your partner use mental math? For which problems? Why?
- Look at your work for Problem 2. Did you rename ones? Tens? Hundreds? How can you tell?
- Explain to a partner how Problem 4 is different than the other problems. What steps did you use to solve this problem?

Exit Ticket (3 minutes)

After the Student Debrief, instruct students to complete the Exit Ticket. A review of their work will help with assessing students' understanding of the concepts that were presented in today's lesson and planning more effectively for future lessons. The questions may be read aloud to the students.

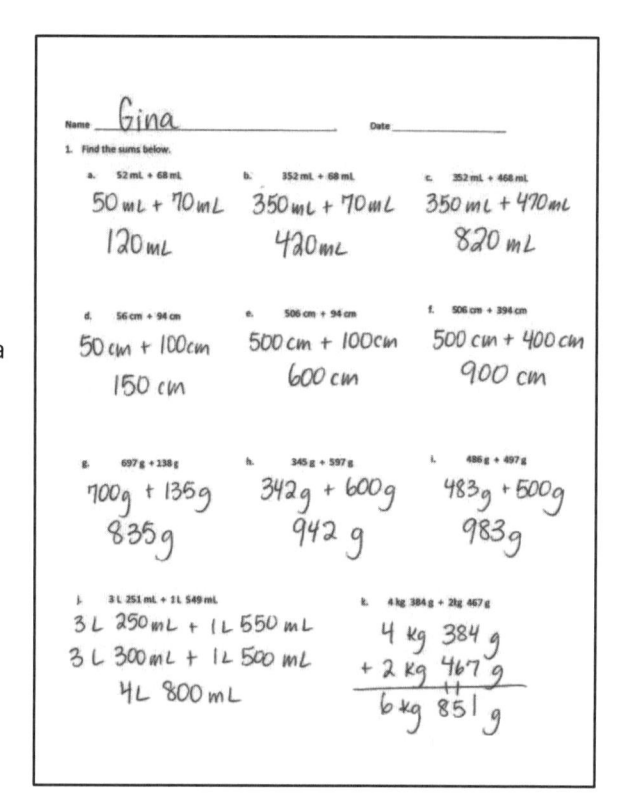

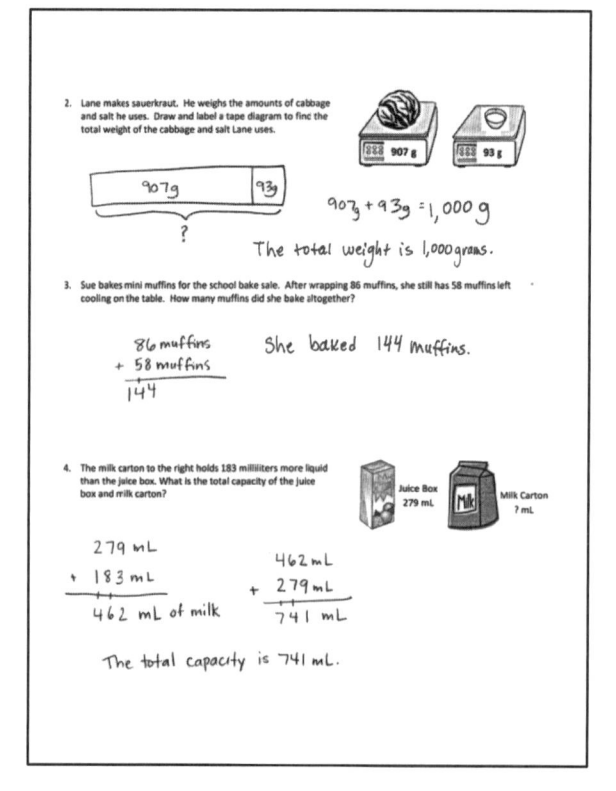

Lesson 16: Add measurements using the standard algorithm to compose larger units twice.

EUREKA MATH

This work is derived from Eureka Math ™ and licensed by Great Minds. ©2015 Great Minds. eureka-math.org

A STORY OF UNITS

Lesson 16 Problem Set 3•2

Name _____ Date _____

1. Find the sums below.

   a. 52 mL + 68 mL

   b. 352 mL + 68 mL

   c. 352 mL + 468 mL

   d. 56 cm + 94 cm

   e. 506 cm + 94 cm

   f. 506 cm + 394 cm

   g. 697 g + 138 g

   h. 345 g + 597 g

   i. 486 g + 497 g

   j. 3 L 251 mL + 1 L 549 mL

   k. 4 kg 384 g + 2 kg 467 g

2. Lane makes sauerkraut. He weighs the amounts of cabbage and salt he uses. Draw and label a tape diagram to find the total weight of the cabbage and salt Lane uses.

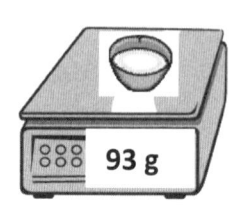

   907 g   93 g

3. Sue bakes mini-muffins for the school bake sale. After wrapping 86 muffins, she still has 58 muffins left cooling on the table. How many muffins did she bake altogether?

4. The milk carton to the right holds 183 milliliters more liquid than the juice box. What is the total capacity of the juice box and milk carton?

   Juice Box
   279 mL

   Milk Carton
   ? mL

A STORY OF UNITS   Lesson 16 Exit Ticket 3•2

Name _____   Date _____

1. Find the sums.

   a. 78 g + 29 g   b. 328 kg + 289 kg   c. 509 L + 293 L

2. The third-grade class sells lemonade to raise funds. After selling 58 liters of lemonade in 1 week, they still have 46 liters of lemonade left. How many liters of lemonade did they have at the beginning?

Lesson 16: Add measurements using the standard algorithm to compose larger units twice.

Name _____   Date _____

1. Find the sums below.

   a. 47 m + 8 m

   b. 47 m + 38 m

   c. 147 m + 383 m

   d. 63 mL + 9 mL

   e. 463 mL + 79 mL

   f. 463 mL + 179 mL

   g. 368 kg + 263 kg

   h. 508 kg + 293 kg

   i. 103 kg + 799 kg

   j. 4 L 342 mL + 2 L 214 mL

   k. 3 kg 296 g + 5 kg 326 g

2. Mrs. Haley roasts a turkey for 55 minutes. She checks it and decides to roast it for an additional 46 minutes. Use a tape diagram to find the total minutes Mrs. Haley roasts the turkey.

3. A miniature horse weighs 268 fewer kilograms than a Shetland pony. Use the table to find the weight of a Shetland pony.

Types of Horses	Weight in kg
Shetland pony	_____ kg
American Saddlebred	478 kg
Clydesdale horse	_____ kg
Miniature horse	56 kg

4. A Clydesdale horse weighs as much as a Shetland pony and an American Saddlebred horse combined. How much does a Clydesdale horse weigh?

Lesson 17

Objective: Estimate sums by rounding and apply to solve measurement word problems.

Suggested Lesson Structure

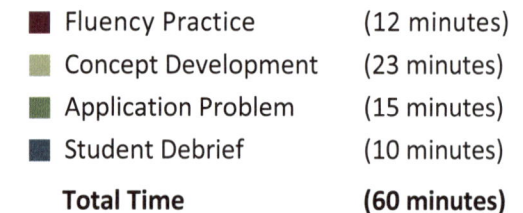

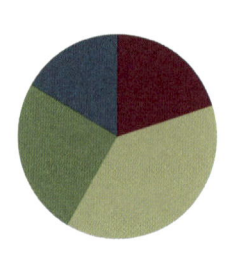

- ■ Fluency Practice   (12 minutes)
- ■ Concept Development   (23 minutes)
- ■ Application Problem   (15 minutes)
- ■ Student Debrief   (10 minutes)
-    **Total Time**   **(60 minutes)**

> **A NOTE ON STANDARDS ALIGNMENT:**
>
> In this lesson, students round to the nearest ten, hundred, and fifty and then analyze the precision of each estimate. When estimating sums, students intentionally make choices that lead to reasonably accurate answers and simple arithmetic. Rounding to the nearest fifty is not part of Grade 3 standards. Its inclusion here combats rigidity in thinking, encouraging students to consider the purpose of their estimates and the degree of accuracy needed rather than simply following procedure. Rounding to the nearest fifty is a Grade 4 standard (**4.NBT.3**) and is not assessed in Grade 3.

Fluency Practice (12 minutes)

- Group Counting   **3.OA.1**   (3 minutes)
- Sprint: Round to the Nearest Ten   **3.NBT.1**   (9 minutes)

Group Counting (3 minutes)

Note: Group counting reviews interpreting multiplication as repeated addition. It reviews foundational strategies for multiplication from Module 1 and anticipates Module 3.

Direct students to count forward and backward, occasionally changing the direction of the count:

- Threes to 30
- Fours to 40
- Sixes to 60
- Sevens to 70
- Eights to 80
- Nines to 90

As students' fluency with skip-counting improves, help them make a connection to multiplication by tracking the number of groups they count using their fingers.

Sprint: Round to the Nearest Ten (9 minutes)

Materials: (S) Round to the Nearest Ten Sprint

Note: This Sprint builds automaticity with rounding skills learned in Lesson 13.

A STORY OF UNITS

Lesson 17 3•2

Concept Development (23 minutes)

Materials: (S) Personal white board

Problem 1: Estimate the sum of 362 + 159 by rounding.   *I do*

- T: What is 362 rounded to the nearest hundred?
- S: 400.
- T: Let's write it directly below 362.
- T: What is 159 rounded to the nearest hundred?
- S: 200.
- T: Let's write it directly below 159.
- T: What is 400 + 200?
- S: 600.
- T: We estimated the sum by rounding to the nearest hundred and got 600.
- T: Let's now round to the nearest ten. (Repeat the process.)
- S: (Find that the sum rounded to the nearest ten is 520.)
- T: We've learned to round to the nearest ten and hundred before. Let's think if there is another way we could round these numbers that would make them easy to add.
- S: They are both really close to a fifty, and those are easy for me to add. → Yeah, 50 + 50 is 100. → You can't round to a fifty! → Why not? Who said so? Makes sense to me. (If no student offers the idea of rounding to the nearest 50, suggest it.)
- T: Okay, let's try it. What is 362 rounded to the nearest fifty?
- S: 350.
- T: 159?
- S: 150.
- T: 350 + 150 is…?
- S: 500.
- T: We have three different estimated sums. Talk to your partner. Without finding the actual sum, which estimate do you think will be closest?
- S: I think rounding to the nearest hundred will be way off. → Me, too. The numbers are pretty far away from the hundred. → Both numbers are close to the halfway point between the hundreds. → Rounding to the nearest ten will be really close because 159 is just 1 away and 362 is just 2 away from our rounded numbers. → Rounding to the fifty will be pretty close, too, but not as close as to the ten because there was a difference of 9 and 12 for both numbers. → And both the numbers were bigger than the 50, too.

> **A NOTE ON STANDARDS ALIGNMENT:**
>
> This problem asks students to round to the nearest fifty, which is part of the Grade 4 standard (**4.NBT.3**).

362 + 159
↓
400 + 200 = 600
360 + 160 = 520
350 + 150 = 500

362 + 159 = 521

Rounding to the nearest ten or fifty gave the most precise estimate this time.

Lesson 17: Estimate sums by rounding and apply to solve measurement word problems.

A STORY OF UNITS   Lesson 17 3•2

T: Let's calculate the actual sum.
S: (Calculate.) It's 521. Wow, rounding to the ten was super close! → Rounding to the fifty was a lot closer than rounding to the hundred. → And, it was easier mental math than rounding to the nearest ten.
T: How did you predict which way of estimating would be closer?
S: We looked at the rounded numbers and thought about how close they were to the actual numbers.
T: We think about how to round in each situation to make our estimates as precise as we need them to be.

Problem 2: Analyze the rounded sums of three expressions with addends close to the halfway point: (A) 349 + 145, (B) 352 + 145, and (C) 352 + 151.

> **NOTES ON ROUNDING PROBLEM 2 EXPRESSIONS:**
> A. 349 + 145 (Numbers round down.)
> B. 352 + 145 (One number rounds up, and one rounds down.)
> C. 352 + 151 (Numbers round up.)

T: (Write the three expressions above on the board.) Take 90 seconds to find the value of these expressions.
S: (Work and check answers.)
T: What do you notice about the sums 494, 497, and 503?
S: They are really close to each other. → They are all between 490 and 510. → The difference between the smallest and greatest is 9.
T: Analyze why the sums were so close by looking at the parts being added. What do you notice?
S: Two of them are exactly the same. → Every problem has a part really close to 350.
T: Let's round each number to the nearest hundred as we did earlier. (Lead students through rounding each addend as pictured below.) Talk to your partner about what you notice.
S: The answers are really different. → The sums were only 9 apart, but the estimates are 200 apart!

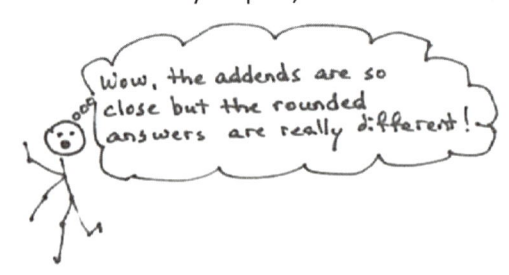

```
349 + 145
300 + 100 = 400

352 + 145
400 + 100 = 500

352 + 151
400 + 200 = 600
```

Wow, the addends are so close but the rounded answers are really different!

T: Why do you think that happened?
S: It's because of how we rounded. → Now I see it. All the numbers we added are really close to the halfway point. → 349 rounded down to 300, but 352 rounded up to 400. → A's numbers both rounded down. For B's numbers, one rounded up and one rounded down. C's numbers both rounded up. → So, in A and C when the numbers rounded the same way, the sums were further away from the actual answer. → B was the closest to the real answer because one went up and one went down.

T: I hear important analysis going on. A very small difference in the numbers can make a difference in the way we round and also make a big difference in the result. How might you get a better estimate when you see that the **addends** are close to halfway between your rounding units?

S: It's like the first problem. We could round to the nearest ten or fifty.

T: That would give us a more precise estimate in cases like these where the numbers are so close to the halfway point.

T: Think about why 352 + 145 had the estimate closest to the precise answer. Share with your partner.

S: It's because one number rounded up and one rounded down. → Yeah, in A and C, either both numbers went down or both went up! → In B, they balanced each other out.

T: Why do we want our estimated sum to be about right?

S: We want to see if our exact answer makes sense. → It also helps with planning, like maybe planning how much to spend at the market. My mom says how much money she has, and we help her make sure we don't spend more.

T: Would all three estimates help you to check if your exact answer is **reasonable**, if it makes sense?

S: No. Only B. → If we used A or C, our exact answer could be way off and we wouldn't know it.

T: So, we need a close estimate to see if our actual sum is reasonable.

Continue with the following possible problem: 253 + 544. Have students estimate by rounding to the nearest ten and fifty to determine which is best for checking whether or not the actual answer is reasonable. To save time, consider dividing the class into two groups; one group rounds to the nearest ten, and the other rounds to the nearest fifty.

Problem 3: Round the sum of 296 + 609. Analyze how rounding to the nearest hundred is nearly the same as rounding to the nearest ten when both addends are close to a hundred.

T: Here is another problem. With your partner, first think about how to round to get the closest answer.

As in Problems 1 and 2, have students analyze the rounded addends before calculating to determine which is best for a precise answer. Then have students calculate the estimated sums, rounding to different units, and compare. Close this problem with an analysis of why this occurred. (Both numbers are very close to the hundreds unit.)

296 + 609

300 + 600 = 900

300 + 610 = 910

296 + 604 = 905

In this case rounding to the hundred or ten was equally precise!

Application Problem (15 minutes)

The doctor prescribed 175 milliliters of medicine on Monday and 256 milliliters of medicine on Tuesday.

a. Estimate how much medicine he prescribed in both days.

b. Precisely how much medicine did he prescribe in both days?

T: To solve Part (a), first determine how you are going to round your numbers.

T: (Allow students time to work the entire problem and possibly to share with a partner. Invite a few students to share with the class how they rounded.)

T: Rounding to the nearest 100 wasn't very precise this time.

> **NOTES ON MULTIPLE MEANS OF ENGAGEMENT:**
>
> Upon evaluating the usefulness of rounding to the nearest ten or hundred, invite students to propose a better method of rounding to check the reasonableness of answers. In this example, rounding one addend to the nearest hundred is a useful strategy.

$175 + 256 = 431$

$200 + 300 = 500$
$180 + 260 = 440$
$200 + 256 = 456$

Rounding to the nearest 100 was easy mental math but not too precise.

Rounding to the nearest 10 was precise, but harder mental math.

Rounding one addend to the nearest 100 is closer than rounding both to 100 and easier mental math than rounding to the nearest 10.

Note: This problem reviews estimating sums to solve word problems, which students learned in today's Concept Development.

Problem Set (10 minutes)

Students should do their personal best to complete the Problem Set within the allotted 10 minutes. For some classes, it may be appropriate to modify the assignment by specifying which problems they work on first. Some problems do not specify a method for solving. Students should solve these problems using the RDW approach used for Application Problems.

> **NOTE ON TIMING:**
>
> The Problem Set in this lesson is allotted 10 minutes. It directly follows the Application Problem, and so the 10 minutes are included within the 15 minutes allotted for the Application Problem.

A STORY OF UNITS
Lesson 17 3•2

Student Debrief (10 minutes)

Lesson Objective: Estimate sums by rounding and apply to solve measurement word problems.

The Student Debrief is intended to invite reflection and active processing of the total lesson experience.

Invite students to review their solutions for the Problem Set. They should check work by comparing answers with a partner before going over answers as a class. Look for misconceptions or misunderstandings that can be addressed in the Debrief. Guide students in a conversation to debrief the Problem Set and process the lesson.

Any combination of the questions below may be used to lead the discussion.

- What were some of your observations about Problem 1(a)? What did the closest estimates have in common?
- Talk to a partner: Which way of rounding in Problem 2 gave an estimate closer to the actual sum?
- How does estimating help you check if your answer is **reasonable**?
- Why might noticing how close the **addends** are to the halfway point change the way you choose to round?
- In Problem 3(a) how did you round? Compare your method with your partner's. Which was closer to the actual answer? Why?
- How did the Application Problem connect to today's lesson?

Exit Ticket (3 minutes)

After the Student Debrief, instruct students to complete the Exit Ticket. A review of their work will help with assessing students' understanding of the concepts that were presented in today's lesson and planning more effectively for future lessons. The questions may be read aloud to the students.

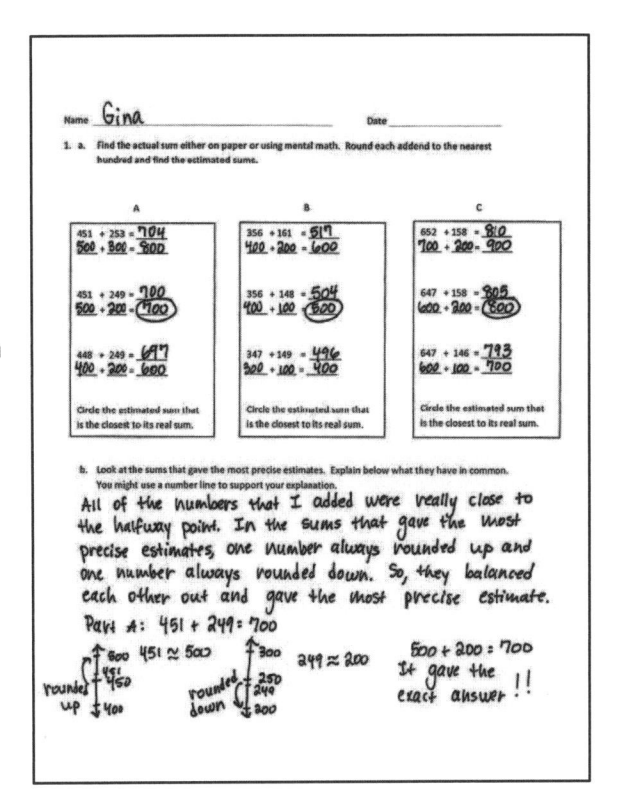

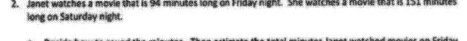

Lesson 17: Estimate sums by rounding and apply to solve measurement word problems.

A

Round to the Nearest Ten

Number Correct: _____

1.	21 ≈		23.	79 ≈	
2.	31 ≈		24.	89 ≈	
3.	41 ≈		25.	99 ≈	
4.	81 ≈		26.	109 ≈	
5.	59 ≈		27.	119 ≈	
6.	49 ≈		28.	149 ≈	
7.	39 ≈		29.	311 ≈	
8.	19 ≈		30.	411 ≈	
9.	36 ≈		31.	519 ≈	
10.	34 ≈		32.	619 ≈	
11.	56 ≈		33.	629 ≈	
12.	54 ≈		34.	639 ≈	
13.	77 ≈		35.	669 ≈	
14.	73 ≈		36.	969 ≈	
15.	68 ≈		37.	979 ≈	
16.	62 ≈		38.	989 ≈	
17.	25 ≈		39.	999 ≈	
18.	35 ≈		40.	1,109 ≈	
19.	45 ≈		41.	1,119 ≈	
20.	75 ≈		42.	3,227 ≈	
21.	85 ≈		43.	5,487 ≈	
22.	15 ≈		44.	7,885 ≈	

A STORY OF UNITS

Lesson 17 Sprint 3•2

B

Round to the Nearest Ten

Number Correct: _____

Improvement: _____

1.	11 ≈	
2.	21 ≈	
3.	31 ≈	
4.	71 ≈	
5.	69 ≈	
6.	59 ≈	
7.	49 ≈	
8.	19 ≈	
9.	26 ≈	
10.	24 ≈	
11.	46 ≈	
12.	44 ≈	
13.	87 ≈	
14.	83 ≈	
15.	78 ≈	
16.	72 ≈	
17.	15 ≈	
18.	25 ≈	
19.	35 ≈	
20.	75 ≈	
21.	85 ≈	
22.	45 ≈	

23.	79 ≈	
24.	89 ≈	
25.	99 ≈	
26.	109 ≈	
27.	119 ≈	
28.	159 ≈	
29.	211 ≈	
30.	311 ≈	
31.	418 ≈	
32.	518 ≈	
33.	528 ≈	
34.	538 ≈	
35.	568 ≈	
36.	968 ≈	
37.	978 ≈	
38.	988 ≈	
39.	998 ≈	
40.	1,108 ≈	
41.	1,118 ≈	
42.	2,337 ≈	
43.	4,578 ≈	
44.	8,785 ≈	

Lesson 17: Estimate sums by rounding and apply to solve measurement word problems.

A STORY OF UNITS

Lesson 17 Problem Set 3•2

Name _____ Date _____

1. a. Find the actual sum either on paper or using mental math. Round each addend to the nearest hundred, and find the estimated sums.

A

451 + 253 = _____
____ + ____ = _____

451 + 249 = _____
____ + ____ = _____

448 + 249 = _____
____ + ____ = _____

Circle the estimated sum that is the closest to its real sum.

B

356 + 161 = _____
____ + ____ = _____

356 + 148 = _____
____ + ____ = _____

347 + 149 = _____
____ + ____ = _____

Circle the estimated sum that is the closest to its real sum.

C

652 + 158 = _____
____ + ____ = _____

647 + 158 = _____
____ + ____ = _____

647 + 146 = _____
____ + ____ = _____

Circle the estimated sum that is the closest to its real sum.

b. Look at the sums that gave the most precise estimates. Explain below what they have in common. You might use a number line to support your explanation.

Lesson 17: Estimate sums by rounding and apply to solve measurement word problems.

2. Janet watched a movie that is 94 minutes long on Friday night. She watched a movie that is 151 minutes long on Saturday night.

   a. Decide how to round the minutes. Then, estimate the total minutes Janet watched movies on Friday and Saturday.

   b. How much time did Janet actually spend watching movies?

   c. Explain whether or not your estimated sum is close to the actual sum. Round in a different way, and see which estimate is closer.

3. Sadie, a bear at the zoo, weighs 182 kilograms. Her cub weighs 74 kilograms.

   a. Estimate the total weight of Sadie and her cub using whatever method you think best.

   b. What is the actual weight of Sadie and her cub? Model the problem with a tape diagram.

Name _____ Date _____

Jesse practices the trumpet for a total of 165 minutes during the first week of school. He practices for 245 minutes during the second week.

a. Estimate the total amount of time Jesse practices by rounding to the nearest 10 minutes.

b. Estimate the total amount of time Jesse practices by rounding to the nearest 100 minutes.

c. Explain why the estimates are so close to each other.

Name _____ Date _____

1. Cathy collects the following information about her dogs, Stella and Oliver.

Stella	
Time Spent Getting a Bath	Weight
36 minutes	32 kg

Oliver	
Time Spent Getting a Bath	Weight
25 minutes	7 kg

Use the information in the charts to answer the questions below.

a. Estimate the total weight of Stella and Oliver.

b. What is the actual total weight of Stella and Oliver?

c. Estimate the total amount of time Cathy spends giving her dogs a bath.

d. What is the actual total time Cathy spends giving her dogs a bath?

e. Explain how estimating helps you check the reasonableness of your answers.

2. Dena reads for 361 minutes during Week 1 of her school's two-week long Read-A-Thon. She reads for 212 minutes during Week 2 of the Read-A-Thon.

   a. Estimate the total amount of time Dena reads during the Read-A-Thon by rounding.

   b. Estimate the total amount of time Dena reads during the Read-A-Thon by rounding in a different way.

   c. Calculate the actual number of minutes that Dena reads during the Read-A-Thon. Which method of rounding was more precise? Why?

A STORY OF UNITS

Mathematics Curriculum

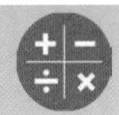

GRADE 3 • MODULE 2

Topic E

Two- and Three-Digit Measurement Subtraction Using the Standard Algorithm

3.NBT.2, 3.NBT.1, 3.MD.1, 3.MD.2

Focus Standard:	3.NBT.2	Fluently add and subtract within 1000 using strategies and algorithms based on place value, properties of operations, and/or the relationship between addition and subtraction.
Instructional Days:	4	
Coherence -Links from:	G2–M2	Addition and Subtraction with Length Units
	G2–M5	Addition and Subtraction Within 1000 with Word Problems to 100
-Links to:	G4–M1	Place Value, Rounding, and Algorithms for Addition and Subtraction

Students work with the standard algorithm for subtraction in Topic E. As in Topic D, they use two- and three-digit metric measurements and intervals of minutes within 1 hour to subtract. The sequence of complexity that builds from Lessons 18–20 mirrors the progression used for teaching addition. In Lesson 18, students begin by decomposing once to subtract, modeling their work on the place value chart. They use three-digit minuends that may contain zeros in the tens or ones place. Students move away from the magnifying glass method used in Grade 2 (see Grade 2, Module 4) but continue to prepare numbers for subtraction by decomposing all necessary digits before performing the operation. By the end of the lesson, they are less reliant on the model of the place value chart and practice using the algorithm with greater confidence.

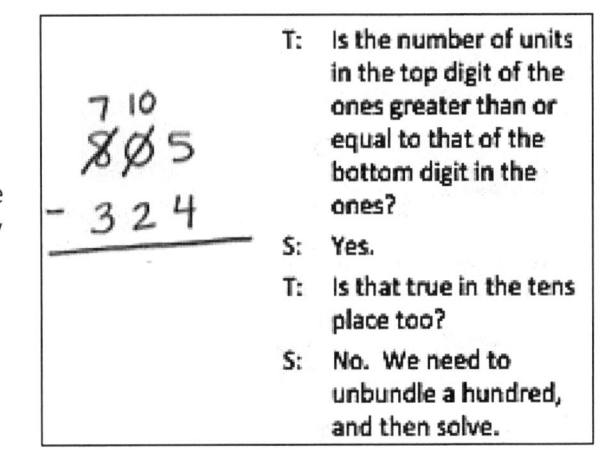

T: Is the number of units in the top digit of the ones greater than or equal to that of the bottom digit in the ones?
S: Yes.
T: Is that true in the tens place too?
S: No. We need to unbundle a hundred, and then solve.

Lesson 19 adds the complexity of decomposing twice to subtract. Minuends may include numbers that contain zeros in the tens *and* ones places. Lesson 20 consolidates the learning from the two prior lessons by engaging students in problem solving with measurements using the subtraction algorithm. As in Lesson 17, students draw to model problems, round to estimate differences, and use the algorithm to subtract precisely. They compare estimates with solutions and assess the reasonableness of their answers.

Lesson 21 synthesizes the skills learned in the second half of the module. Students round to estimate the sums and differences of measurements in word problem contexts. They draw to model problems and apply the algorithms to solve each case introduced in Topics D and E precisely. As in previous lessons, students use their estimates to reason about their solutions.

T: Is the number of units in the top digit of the ones greater than or equal to that of the bottom digit in the ones?

S: No. We need to unbundle a ten.

T: How about in the tens place?

S: No. We need to unbundle a hundred too. Then we can solve.

A Teaching Sequence Toward Mastery of Two- and Three-Digit Measurement Subtraction Using the Standard Algorithm

Objective 1: Decompose once to subtract measurements including three-digit minuends with zeros in the tens or ones place.
(Lesson 18)

Objective 2: Decompose twice to subtract measurements including three-digit minuends with zeros in the tens and ones places.
(Lesson 19)

Objective 3: Estimate differences by rounding and apply to solve measurement word problems.
(Lesson 20)

Objective 4: Estimate sums and differences of measurements by rounding, and then solve mixed word problems.
(Lesson 21)

Lesson 18

Objective: Decompose once to subtract measurements including three-digit minuends with zeros in the tens or ones place.

Suggested Lesson Structure

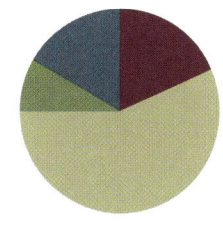

- ■ Fluency Practice   (11 minutes)
- ■ Application Problem (5 minutes)
- ■ Concept Development (34 minutes)
- ■ Student Debrief   (10 minutes)
- **Total Time**   **(60 minutes)**

Fluency Practice (11 minutes)

- Group Counting   **3.OA.1**   (3 minutes)
- Subtract Mentally **3.NBT.2**   (4 minutes)
- Estimate and Add **3.NBT.2**   (4 minutes)

Group Counting (3 minutes)

Note: Group counting reviews interpreting multiplication as repeated addition. It reviews foundational strategies for multiplication from Module 1 and anticipates Module 3.

Direct students to count forward and backward, occasionally changing the direction of the count:

- Threes to 30
- Fours to 40
- Sixes to 60
- Sevens to 70
- Eights to 80
- Nines to 90

As students' fluency with skip-counting improves, help them make a connection to multiplication by tracking the number of groups they count using their fingers.

Subtract Mentally (4 minutes)

Note: This activity anticipates the role of place value in the subtraction algorithm.

T: (Write 10 – 3 = ___.) Say the number sentence in units of one.
S: 10 ones – 3 ones = 7 ones.

A STORY OF UNITS   Lesson 18   3•2

Continue with the following sequence: 11 – 3 and 61 – 3 (as pictured below at right).

- T: (Write 100 – 30 = ___.)
  Now say the number sentences in units of ten.
- T: 10 tens – 3 tens = 7 tens.

| 10 – 3 = 7 | 11 – 3 = 8 | 61 – 3 = 58 |
| 100 – 30 = 70 | 110 – 30 = 80 | 610 – 30 = 580 |

Continue with the following sequence: 110 – 30 and 610 – 30.

Repeat with the following possible sequences:

- 10 – 5, 12 – 5, and 73 – 5
- 100 – 50, 120 – 50, and 730 – 50

Estimate and Add (4 minutes)

Materials: (S) Personal white board

Note: This activity reviews rounding to estimate sums from Lesson 17.

- T: (Write 38 + 23 ≈ ___.) Say the addition problem.
- S: 38 + 23.
- T: Give me the new addition problem if we round each number to the nearest ten.
- S: 40 + 20.
- T: (Write 38 + 23 ≈ 40 + 20.) What's 40 + 20?
- S: 60.
- T: So, 38 + 23 should be close to …?
- S: 60.
- T: On your personal white board, solve 38 + 23.
- S: (Solve.)

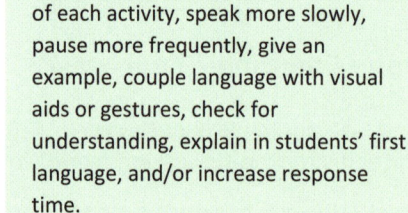

> **NOTES ON MULTIPLE MEANS OF ACTION AND EXPRESSION:**
>
> Fluency activities are fun, fast-paced math games, but don't leave English language learners behind. At the start of each activity, speak more slowly, pause more frequently, give an example, couple language with visual aids or gestures, check for understanding, explain in students' first language, and/or increase response time.

Continue with the following possible sequence: 24 + 59, 173 + 49, and 519 + 185.

Application Problem (5 minutes)

Tara brings 2 bottles of water on her hike. The first bottle has 471 milliliters of water, and the second bottle has 354 milliliters of water. How many milliliters of water does Tara bring on her hike?

Note: This problem reviews composing units once to add. It will be used to reintroduce the place value chart during Part 1 of the Concept Development.

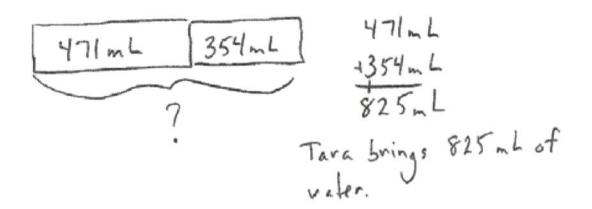

Lesson 18: Decompose once to subtract measurements including three-digit minuends with zeros in the tens or ones place.

A STORY OF UNITS   Lesson 18 3•2

Concept Development (34 minutes)

Materials: (T) Unlabeled place value chart (Lesson 14 Template) (S) Personal white board, unlabeled place value chart (Lesson 14 Template)

Part 1: Use the place value chart to model decomposing once to subtract with three-digit minuends.

Students start with the unlabeled place value chart template in their personal white boards.

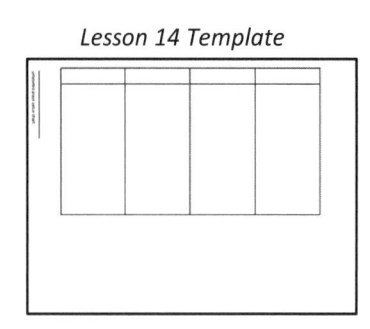

- T: Tara has 132 milliliters of water left after hiking. How can we find out how many milliliters of water Tara drinks while she is hiking?
- S: We can subtract. → We can subtract 132 milliliters from 825 milliliters. → She drank 825 milliliters – 132 milliliters.
- T: Let's write that vertically in the workspace below the place value chart on our personal white board and then model the problem on our place value charts. (Model writing 825 – 132 as a vertical problem.) On your place value chart, draw place value disks to represent the amount of water Tara starts with.
- S: (Draw place value disks.) 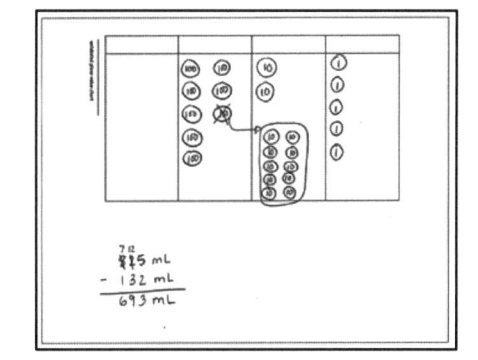
- T: Let's get ready to subtract. Look at your vertical subtraction problem. How many ones do we need to subtract from the 5 ones that are there now?
- S: 2 ones.
- T: Can we subtract 2 ones from 5 ones?
- S: Yes!
- T: How many tens are we subtracting from 2 tens?
- S: 3 tens.
- T: Can we subtract 3 tens from 2 tens?
- S: No!
- T: Why not?
- S: There aren't enough tens to subtract from. → 3 tens is more than 2 tens.
- T: To get more tens so that we can subtract, we have to unbundle 1 hundred into tens. How many tens in 1 hundred?
- S: 10 tens!
- T: (Model the process of unbundling 1 hundred into 10 tens, as shown to the right. Have students work along with you.) To start off, we had 8 hundreds and 2 tens. Now, how many hundreds and tens do we have?
- S: 7 hundreds and 12 tens!
- T: Now that we have 12 tens, can we take 3 tens away?
- S: Yes!

Lesson 14 Template

Lesson 18: Decompose once to subtract measurements including three-digit minuends with zeros in the tens or ones place.

A STORY OF UNITS   Lesson 18 3•2

T: Now let's move to the hundreds place. Can we subtract 1 hundred from 7 hundreds?
S: Yes!
T: We're ready to subtract. Cross off the ones, tens, and hundreds that are being subtracted. (Model as students work along.)
T: So, what's the result?
S: 693.
T: So, that's it. Our answer is 693?
S: No! We were looking for the amount of water, not just a number. It's 693 milliliters!
T: Answer the question with a full statement.
S: Tara drank 693 milliliters of water on her hike.

MP.2

> **NOTES ON MULTIPLE MEANS OF REPRESENTATION:**
>
> Use color to customize the presentation of decomposing to subtract. Enhance learners' perception of the information by consistently displaying hundreds in one color (e.g., red), while displaying tens in a different color (e.g., green). Consider varying the colors for each place value unit when teaching the standard algorithm from day to day so that students continue to look for a value, rather than for a color.

Continue with the following suggested sequence:

- 785 cm – 36 cm
- 440 g – 223 g
- 508 mL – 225 mL

Part 2: Subtract using the standard algorithm.

Write or project the following problem:

Nooran buys 507 grams of grapes at the market on Tuesday. On Thursday, he buys 345 grams of grapes. How many more grams of grapes did Nooran buy on Tuesday than on Thursday?

T: Let's model this problem with a tape diagram to figure out what we need to do to solve. Draw with me on your board. (Model.) How should we solve this problem?
S: We can subtract, 507 grams – 345 grams. → We're looking for the part that's different so we subtract. → To find a missing part, subtract.
T: Write the equation, and then talk to your partner. Is this problem easily solved using mental math? Why or why not?
S: Not really. → It's easy to subtract 300 from 500, but the 7 and the 45 aren't very friendly.
T: Like with addition problems that aren't easily solved with simplifying strategies, we can use the standard algorithm to solve subtraction problems that aren't easily solved with simplifying strategies. Rewrite the problem vertically on your board if you need to.
S: (Rewrite problem vertically.)
T: Before we subtract, let's see if any unbundling needs to be done. Are there enough ones to subtract 5 ones?
S: Yes.
T: Are there enough tens to subtract 4 tens?
S: No, 0 tens is less than 4 tens.
T: How can we get some more tens?
S: We can go to the hundreds place. → We can unbundle 1 hundred to make 10 tens.

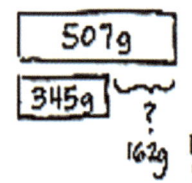

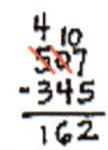

Lesson 18: Decompose once to subtract measurements including three-digit minuends with zeros in the tens or ones place.

T: How many hundreds are in the number on top?
S: 5 hundreds.
T: When we unbundle 1 hundred to make 10 tens, how many hundreds and tens will the top number have?
S: 4 hundreds and 10 tens.
T: (Model.) Do we have enough hundreds to subtract 3 hundreds?
S: Yes.
T: We are ready to subtract! Solve the problem on your board.
T: (Model as shown on the previous page.) How many more grams of grapes did Nooran buy on Tuesday?
S: 162 more grams of grapes!
T: Label the unknown on your tape diagram with the answer.

Continue with the following suggested sequence. Students should unbundle all necessary digits before performing the operation.

- 513 cm − 241 cm
- 760 g − 546 g
- 506 mL − 435 mL

Problem Set (10 minutes)

Students should do their personal best to complete the Problem Set within the allotted 10 minutes. For some classes, it may be appropriate to modify the assignment by specifying which problems they work on first. Some problems do not specify a method for solving. Students should solve these problems using the RDW approach used for Application Problems.

NOTES ON THE PROBLEM SET:

The problems on the Problem Set are written horizontally so that students do not assume that they need to use the standard algorithm to solve. Mental math may be a more efficient strategy in some cases. Invite students to use the algorithm as a strategic tool, purposefully choosing it rather than defaulting to it.

A STORY OF UNITS

Lesson 18 3•2

Student Debrief (10 minutes)

Lesson Objective: Decompose once to subtract measurements including three-digit minuends with zeros in the tens or ones place.

The Student Debrief is intended to invite reflection and active processing of the total lesson experience.

Invite students to review their solutions for the Problem Set. They should check work by comparing answers with a partner before going over answers as a class. Look for misconceptions or misunderstandings that can be addressed in the Debrief. Guide students in a conversation to debrief the Problem Set and process the lesson.

Any combination of the questions below may be used to lead the discussion.

- What is the relationship between Problems 1(a), 1(b), and 1(c)?
- How are Problems 1(j) and 1(k) different from the problems that come before them?
- Invite students to share the tape diagram used to solve Problem 2.
- Compare Problems 2 and 4. What extra step was needed to solve Problem 4? What models could be used to solve this problem?
- Describe the steps of the standard algorithm for subtraction.

Exit Ticket (3 minutes)

After the Student Debrief, instruct students to complete the Exit Ticket. A review of their work will help with assessing students' understanding of the concepts that were presented in today's lesson and planning more effectively for future lessons. The questions may be read aloud to the students.

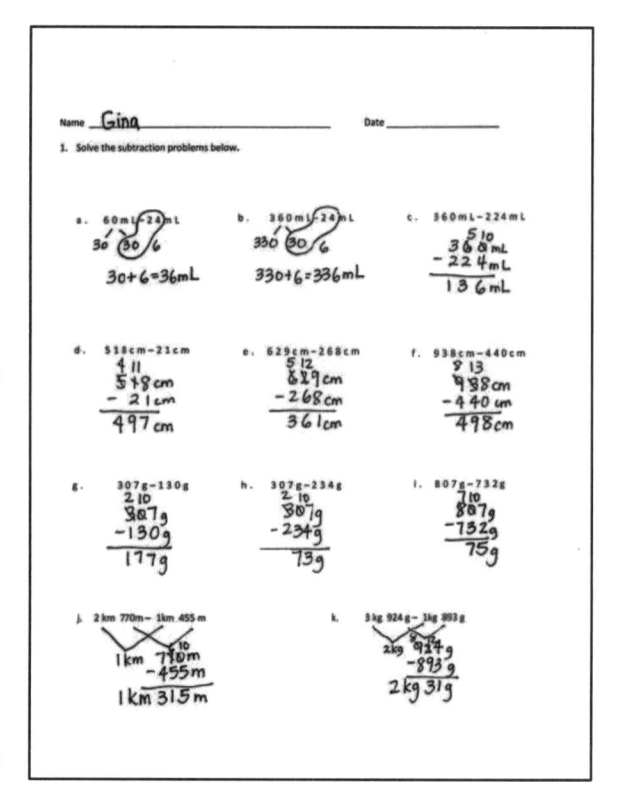

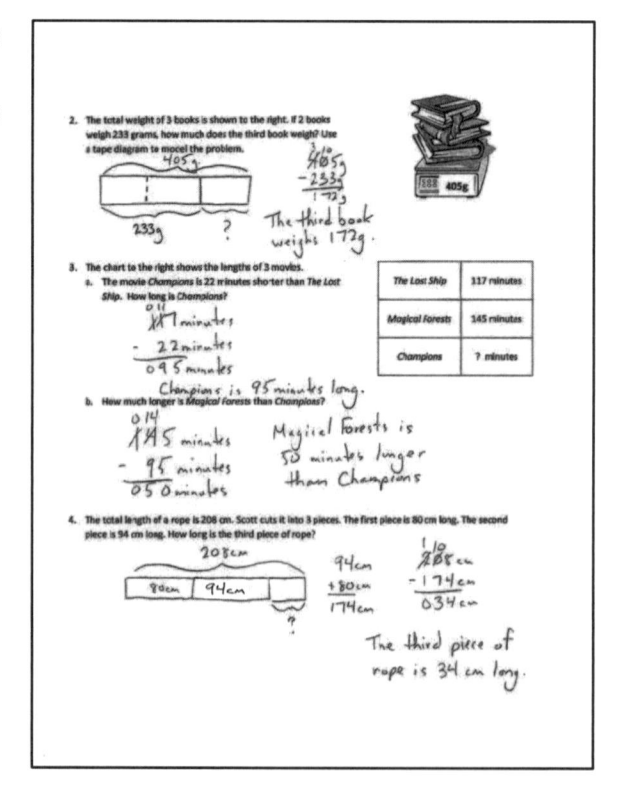

226 Lesson 18: Decompose once to subtract measurements including three-digit minuends with zeros in the tens or ones place.

EUREKA MATH

This work is derived from Eureka Math™ and licensed by Great Minds. ©2015 Great Minds. eureka-math.org

Name _____ Date _____

1. Solve the subtraction problems below.

   a. 60 mL − 24 mL

   b. 360 mL − 24 mL

   c. 360 mL − 224 mL

   d. 518 cm − 21 cm

   e. 629 cm − 268 cm

   f. 938 cm − 440 cm

   g. 307 g − 130 g

   h. 307 g − 234 g

   i. 807 g − 732 g

   j. 2 km 770 m − 1 km 455 m

   k. 3 kg 924 g − 1 kg 893 g

2. The total weight of 3 books is shown to the right. If 2 books weigh 233 grams, how much does the third book weigh? Use a tape diagram to model the problem.

3. The chart to the right shows the lengths of three movies.

   a. The movie *Champions* is 22 minutes shorter than *The Lost Ship*. How long is *Champions*?

The Lost Ship	117 minutes
Magical Forests	145 minutes
Champions	? minutes

   b. How much longer is *Magical Forests* than *Champions*?

4. The total length of a rope is 208 centimeters. Scott cuts it into 3 pieces. The first piece is 80 centimeters long. The second piece is 94 centimeters long. How long is the third piece of rope?

Name _____ Date _____

1. Solve the subtraction problems below.

   a. 381 mL – 146 mL

   b. 730 m – 426 m

   c. 509 kg – 384 kg

2. The total length of a banner is 408 centimeters. Carly paints it in 3 sections. The first 2 sections she paints are 187 centimeters long altogether. How long is the third section?

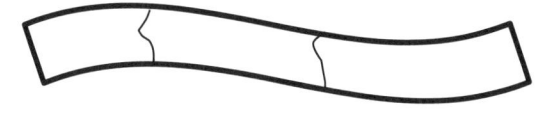

Name _____   Date _____

1. Solve the subtraction problems below.

   a. 70 L – 46 L

   b. 370 L – 46 L

   c. 370 L – 146 L

   d. 607 cm – 32 cm

   e. 592 cm – 258 cm

   f. 918 cm – 553 cm

   g. 763 g – 82 g

   h. 803 g – 542 g

   i. 572 km – 266 km

   j. 837 km – 645 km

A STORY OF UNITS

Lesson 18 Homework 3•2

2. The magazine weighs 280 grams less than the newspaper. The weight of the newspaper is shown below. How much does the magazine weigh? Use a tape diagram to model your thinking.

   454 g

3. The chart to the right shows how long it takes to play 3 games.

   a. Francesca's basketball game is 22 minutes shorter than Lucas's baseball game. How long is Francesca's basketball game?

Lucas's Baseball Game	180 minutes
Joey's Football Game	139 minutes
Francesca's Basketball Game	? minutes

   b. How much longer is Francesca's basketball game than Joey's football game?

Lesson 19

Objective: Decompose twice to subtract measurements including three-digit minuends with zeros in the tens and ones places.

Suggested Lesson Structure

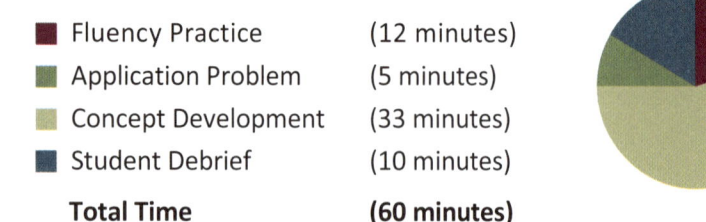

- ■ Fluency Practice   (12 minutes)
- ■ Application Problem   (5 minutes)
- ■ Concept Development   (33 minutes)
- ■ Student Debrief   (10 minutes)
- **Total Time**   **(60 minutes)**

Fluency Practice (12 minutes)

- Subtract Mentally **3.NBT.2**   (4 minutes)
- Use Subtraction Algorithm with Measurements **3.MD.2**   (4 minutes)
- Round Three- and Four-Digit Numbers **3.NBT.1**   (4 minutes)

Subtract Mentally (4 minutes)

Note: This activity emphasizes the role of place value in the subtraction algorithm.

    T: (Write 10 – 5 = ___.) Say the number sentence in units of one.
    S: 10 ones – 5 ones = 5 ones.

Repeat the process outlined in Lesson 18. Use the following suggested sequence: 12 ones – 5 ones, 42 ones – 5 ones, 10 tens – 5 tens, 12 tens – 5 tens, and 42 tens – 5 tens.

Use Subtraction Algorithm with Measurements (4 minutes)

Materials: (S) Personal white board

Note: This activity reviews the role of place value in the subtraction algorithm from Lesson 18.

    T: (Write 80 L – 26 L = ___.) On your personal white board, solve using the standard algorithm.

Continue with the following possible sequence: 380 L – 26 L, 380 L – 126 L, 908 g – 25 g, and 908 g – 425 g.

A STORY OF UNITS   Lesson 19   3•2

Round Three- and Four-Digit Numbers (4 minutes)

Materials: (S) Personal white board

Note: This activity reviews rounding to the nearest hundred from Lesson 14.

T: (Write 253 ≈ ___.) What is 253 rounded to the nearest hundred?
S: 300.

Repeat the process outlined in Lesson 15, rounding numbers only to the nearest hundred. Use the following possible suggestions: 253; 1,253; 735; 1,735; 850; 1,850; 952; 1,371; and 1,450.

Application Problem (5 minutes)

Jolene brings an apple and an orange with her to school. The weight of both pieces of fruit together is 417 grams. The apple weighs 223 grams. What is the weight of Jolene's orange?

[Student work shown: tape diagram with 417g total, 223g and ? parts; vertical subtraction 417 − 223 = 194g with decomposition marks showing 3 and 11; "Jolene's orange weighs 194 grams."]

Note: This problem reviews unbundling once to subtract. It also provides a context leading into the Concept Development.

Concept Development (33 minutes)

Materials: (S) Personal white board

Part 1: Decompose twice using the standard algorithm for subtraction.

T: In the Application Problem, Jolene's apple weighs 223 grams and her orange weighs 194 grams. (Draw or project the tape diagrams shown at right.) What does the question mark in these tape diagrams represent?

S: How much heavier the apple is than the orange.
→ How much more the apple weighs, in grams.

T: Tell a partner what expression you can use to find out how much heavier the apple is than the orange. Write the problem vertically on your personal white board.

S: (Write problem vertically.)

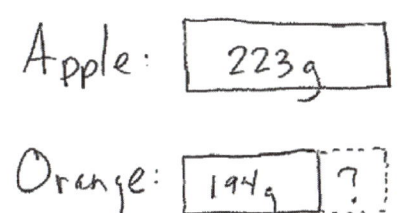

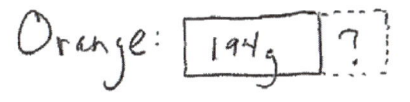

[Diagram at right: Apple: 223g; Orange: 194g ?]

Lesson 19: Decompose twice to subtract measurements including three-digit minuends with zeros in the tens and ones places.

A STORY OF UNITS   Lesson 19 3•2

T: Before we subtract, what needs to be done?
S: We need to make sure we can subtract each place. → We have to see if any tens or hundreds need to be unbundled.
T: Do we have enough ones to subtract?
S: No. We need to change 1 ten for 10 ones.
T: How about in the tens place?
S: No. We also need to change 1 hundred for 10 tens. Then, we can solve.
T: Unbundle or change the ten. How many tens and ones do we have now?
S: 1 ten and 13 ones.
T: Unbundle or change the hundred. How many hundreds and tens do we have now?
S: 1 hundred and 11 tens.
T: Are we ready to subtract?
S: Yes!
T: Solve the problem on your board.
S: (Solve as shown to the right.)
T: How much heavier is the apple than the orange?
S: The apple is 29 grams heavier than the orange!

> **NOTES ON MULTIPLE MEANS OF REPRESENTATION:**
>
> Use color to customize the presentation of the tape diagram. Displaying a green bar for the apple and an orange bar for the orange may enhance learners' perception of the information.
>
> Alternatively, students may value a vertical tape diagram if it gives them a better sense of heavy and less heavy.

Subtraction Complete

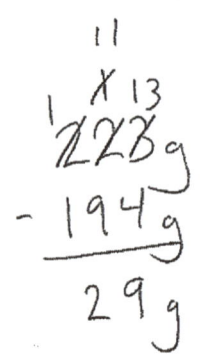

$$\begin{array}{r} \overset{1}{\cancel{2}}\overset{1}{\cancel{2}}\overset{13}{\cancel{3}}g \\ -194\,g \\ \hline 29\,g \end{array}$$

Continue with the following suggested sequence. Students should prepare their problems for subtraction by unbundling all necessary digits before performing the operation.

- 342 cm – 55 cm
- 764 g – 485 g
- 573 mL – 375 mL

T: How are the subtraction problems we've solved so far different from those we solved yesterday?
S: Yesterday, we only had to unbundle once. Today, we had to unbundle twice.

Part 2: Use the standard algorithm to subtract three-digit numbers with zeros in various positions.

Write or project the following problem:

Kerrin has 703 milliliters of water in a pitcher. She pours some water out. Now, 124 milliliters are left in the pitcher. How much water did Kerrin pour out?

T: Let's solve this problem using the algorithm. What needs to be done first?
S: We need to unbundle a ten.
T: What digit is in the tens place on top?

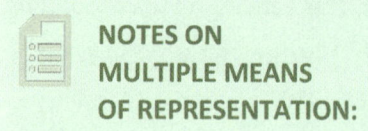

Lesson 19: Decompose twice to subtract measurements including three-digit minuends with zeros in the tens and ones places.

A STORY OF UNITS   Lesson 19 3•2

S: Zero.
T: Can we unbundle 0 tens?
S: No!
T: Where can we get tens?
S: We can change 1 hundred into 10 tens!
T: Change the hundred into tens on your board. (Model.) How many hundreds and tens does the top number have now?
S: 6 hundreds and 10 tens.
T: Why aren't we ready to subtract yet?
S: We still have to change 1 ten for 10 ones.
T: Finish unbundling on your board and complete the subtraction. (Model.) How many milliliters of water did Kerrin pour out?
S: She poured out 579 milliliters of water!

Continue with the following suggested sequence. Students should prepare their problems for subtraction by unbundling all necessary digits before performing the operation.

- 703 cm – 37 cm
- 700 mL – 356 mL
- 500 g – 467 g

Problem Set (10 minutes)

Students should do their personal best to complete the Problem Set within the allotted 10 minutes. For some classes, it may be appropriate to modify the assignment by specifying which problems they work on first. Some problems do not specify a method for solving. Students should solve these problems using the RDW approach used for Application Problems.

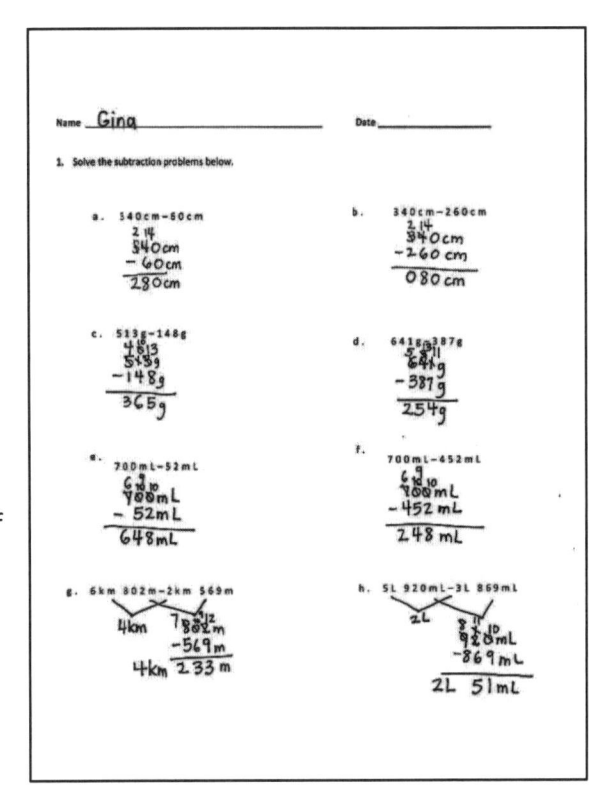

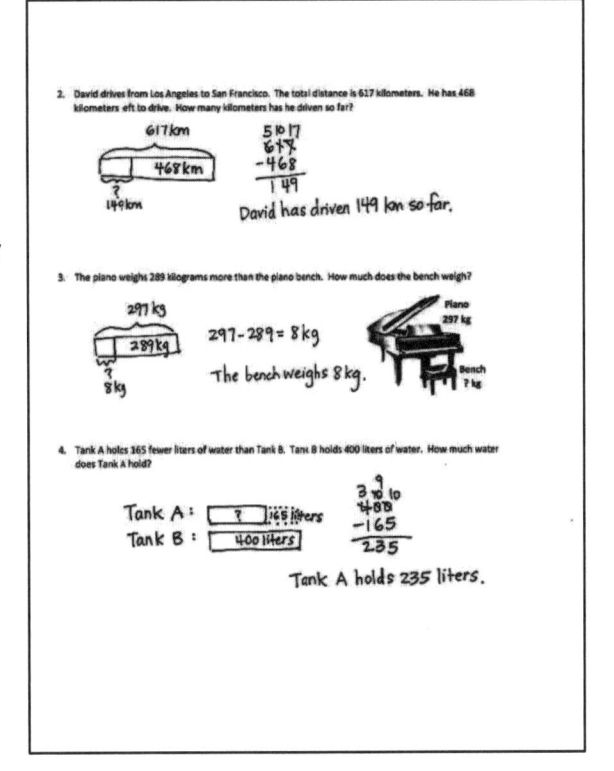

Lesson 19: Decompose twice to subtract measurements including three-digit minuends with zeros in the tens and ones places.

Student Debrief (10 minutes)

Lesson Objective: Decompose twice to subtract measurements including three-digit minuends with zeros in the tens and ones places.

The Student Debrief is intended to invite reflection and active processing of the total lesson experience.

Invite students to review their solutions for the Problem Set. They should check work by comparing answers with a partner before going over answers as a class. Look for misconceptions or misunderstandings that can be addressed in the Debrief. Guide students in a conversation to debrief the Problem Set and process the lesson.

Any combination of the questions below may be used to lead the discussion.

- Which strategy did you use to solve Problem 1(a)? Why? (Students may want to talk about subtracting 6 tens from 34 tens rather than decomposing.)
- Invite students to articulate the steps they followed to solve Problem 4.
- Why is it important to unbundle or change all of your units before subtracting?

Exit Ticket (3 minutes)

After the Student Debrief, instruct students to complete the Exit Ticket. A review of their work will help with assessing students' understanding of the concepts that were presented in today's lesson and planning more effectively for future lessons. The questions may be read aloud to the students.

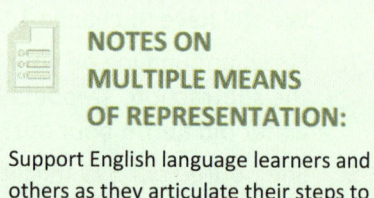

> **NOTES ON MULTIPLE MEANS OF REPRESENTATION:**
>
> Support English language learners and others as they articulate their steps to solve Problem 4. Give students the choice of explaining in their first language. Making this a partner–share activity may relieve students of anxiety in front of a large group. Some students may benefit from sentence starters, such as, "First, I read ____. Then, I drew ____. Next, I labeled ____. Then, I wrote my equation: _____. Last, I wrote my answer statement, which was _____."

Name _____ Date _____

1. Solve the subtraction problems below.

   a. 340 cm − 60 cm

   b. 340 cm − 260 cm

   c. 513 g − 148 g

   d. 641 g − 387 g

   e. 700 mL − 52 mL

   f. 700 mL − 452 mL

   g. 6 km 802 m − 2 km 569 m

   h. 5 L 920 mL − 3 L 869 mL

2. David is driving from Los Angeles to San Francisco. The total distance is 617 kilometers. He has 468 kilometers left to drive. How many kilometers has he driven so far?

3. The piano weighs 289 kilograms more than the piano bench. How much does the bench weigh?

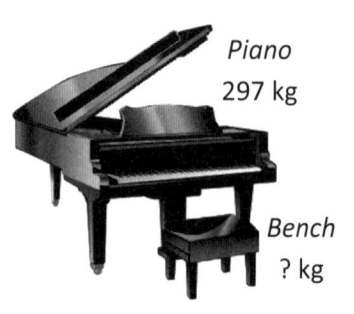

Piano
297 kg

Bench
? kg

4. Tank A holds 165 fewer liters of water than Tank B. Tank B holds 400 liters of water. How much water does Tank A hold?

A STORY OF UNITS Lesson 19 Exit Ticket 3•2

Name _____ Date _____

1. Solve the subtraction problems below.

   a. 346 m – 187 m

   b. 700 kg – 592 kg

2. The farmer's sheep weighs 647 kilograms less than the farmer's cow. The cow weighs 725 kilograms. How much does the sheep weigh?

A STORY OF UNITS

Lesson 19 Homework 3•2

Name _____ Date _____

1. Solve the subtraction problems below.

   a. 280 g – 90 g

   b. 450 g – 284 g

   c. 423 cm – 136 cm

   d. 567 cm – 246 cm

   e. 900 g – 58 g

   f. 900 g – 358 g

   g. 4 L 710 mL – 2 L 690 mL

   h. 8 L 830 mL – 4 L 378 mL

2. The total weight of a giraffe and her calf is 904 kilograms. How much does the calf weigh? Use a tape diagram to model your thinking.

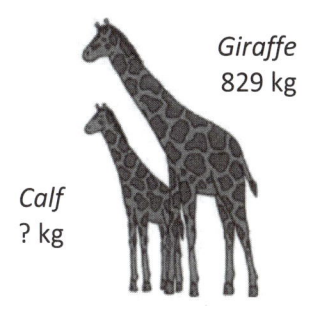

Giraffe
829 kg

Calf
? kg

3. The Erie Canal runs 584 kilometers from Albany to Buffalo. Salvador travels on the canal from Albany. He must travel 396 kilometers more before he reaches Buffalo. How many kilometers has he traveled so far?

4. Mr. Nguyen fills two inflatable pools. The kiddie pool holds 185 liters of water. The larger pool holds 600 liters of water. How much more water does the larger pool hold than the kiddie pool?

A STORY OF UNITS

Lesson 20   3•2

Lesson 20

Objective: Estimate differences by rounding and apply to solve measurement word problems.

Suggested Lesson Structure

- Fluency Practice   (12 minutes)
- Concept Development   (23 minutes)
- Application Problem   (15 minutes)
- Student Debrief   (10 minutes)
  Total Time   **(60 minutes)**

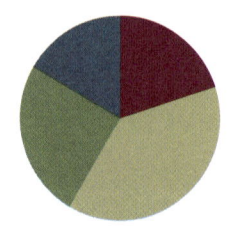

Fluency Practice (12 minutes)

- Sprint: Round to the Nearest Hundred **3.NBT.1**   (9 minutes)
- Use Subtraction Algorithm with Measurements **3.MD.2** (3 minutes)

Sprint: Round to the Nearest Hundred (9 minutes)

Materials: (S) Round to the Nearest Hundred Sprint

Note: This activity builds automaticity with rounding to the nearest hundred from Lesson 14.

Use Subtraction Algorithm with Measurements (3 minutes)

Materials: (S) Personal white board

Note: This activity reviews the standard algorithm taught in Module 2.

   T: (Write 50 L – 28 L = ___.) On your personal white board, solve using the standard algorithm.

Repeat the process outlined in Lesson 19 using the following suggested sequence: 50 L – 28 L, 450 L – 28 L, 450 L – 228 L, 604 g – 32 g, and 604 g – 132 g.

A STORY OF UNITS   Lesson 20   3•2

Concept Development (23 minutes)

Materials: (S) Personal white board

Problem 1: Estimate 362 – 189 by rounding.

T: What is 362 rounded to the nearest hundred?

S: 400.

T: Let's write it directly below 362. (Allow students time to write 400 below 362.) What is 189 rounded to the nearest hundred?

S: 200.

T: Let's write it directly below 189. (Allow students time to write 200 below 189.) What is 400 – 200?

```
362 - 189
    ↓
400 - 200 = 200
360 - 190 = 170
362 - 200 = 162

362 - 189 = 173
```

Rounding both numbers to the nearest ten was closest but rounding the known part to the nearest hundred was close AND easy mental math!

S: 200.

T: We estimated the difference by rounding both numbers to the nearest hundred and got 200. Let's now find the difference by rounding to the nearest ten.

Repeat the process. Students find that the difference rounded to the nearest ten is 360 – 190 or 36 tens – 19 tens, which is 170 or 17 tens.

T: We rounded to the nearest ten and hundred. Is there another easy way we could round these numbers so it is easy to subtract?

S: Since we're subtracting, 362 is the whole and 189 is a part. Let's just round the part. 362 – 200 is 162. That is really easy mental math.

T: So, round only 189 because it is already so close to 200?

S: Yeah. The answer is closer than if we round the whole because 362 isn't very close to 400.

T: How about 380 – 180 = 200? What are our rounded answers?

S: 200, 170, and 162.

T: Let's see which is closest. Solve the problem and discuss which rounded solution is closest. (Solve and discuss.) Rounding to the nearest ten was the closest answer. Which was the easiest mental math?

S: Rounding to the hundred. → Yes, but the answer was much closer when we rounded the known part, like we did with 362 – 200, and it was still easy.

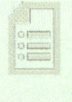

> **NOTES ON MULTIPLE MEANS OF ACTION AND EXPRESSION:**
>
> Learners differ in their internal organization and working memory. Facilitate students' use of the standard algorithm as a desktop checklist. Consider including the following:
>
> 1. Whisper read the problem.
> 2. Say the larger number in unit form.
> 3. Scan. Are you ready to subtract?
> 4. Record bundling and unbundling.
> 5. Check your answer with your estimate, place value disks, or partner.
> 6. Correct any mistake.

EUREKA MATH™   Lesson 20:   Estimate differences by rounding and apply to solve measurement word problems.

A STORY OF UNITS   Lesson 20   3•2

T: Rounding the known part we are subtracting from the whole is easy, and in this case, gave us a pretty good estimate. How does comparing your actual answer with your estimation help you check your calculation?

S: We saw that our answer was not crazy. → If the estimate is really different than the real answer, we can see that we might've made a mistake.

T: Rounding to estimate is a tool that helps us simplify calculations to help us make sure our actual answers are reasonable. Rounding can also be useful when we don't need an exact answer. I know it isn't as precise as an actual calculation, but sometimes an idea of an amount is all the information I need.

Problem 2: Analyze the estimated differences of four expressions with subtrahends close to the halfway point: (A) 349 – 154, (B) 349 – 149, (C) 351 – 154, and (D) 351 – 149.

T: (Write the four expressions above on the board.) Take 90 seconds to find the value of these expressions.

S: (Work and check answers.)

T: What do you notice about the differences: 195, 200, 197, and 202?

S: They are really close to each other. → They are all between 195 and 202. → The difference between the smallest and greatest is 7.

T: Analyze why the differences were so close by looking at the totals and the parts being subtracted. What do you notice?

S: Two of them are exactly the same. → The totals are all really close to 350. → The part being subtracted is really close to 150.

T: Let's round to the nearest hundred as we did earlier. (Lead students through rounding each number to the nearest hundred and finding the differences.)

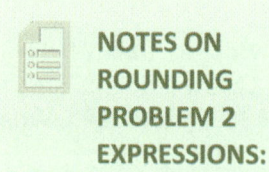

> **NOTES ON ROUNDING PROBLEM 2 EXPRESSIONS:**
>
> A. 349 – 154
> (Whole rounds down. Part rounds up.)
>
> B. 349 – 149
> (Whole rounds down. Part rounds down.)
>
> C. 351 – 154
> (Whole rounds up. Part rounds up.)
>
> D. 351 – 149
> (Whole rounds up. Part rounds down.)

MP.6

349 – 154
300 – 200 = 100

349 – 149
300 – 100 = 200

351 – 154
400 – 200 = 200

351 – 149
400 – 100 = 300

Wow, the whole and the known part are so close but the answers are REALLY different!!

S: The answers came out really different!

Lesson 20: Estimate differences by rounding and apply to solve measurement word problems.

A STORY OF UNITS   Lesson 20   3•2

T: Analyze the rounding with your partner. Did we round up or down?
S: In A, the total rounded down and the part rounded up. → In B, they both rounded down. → In C, they both rounded up. → In D, the total rounded up and the part rounded down.
T: Which estimates are closest to the actual answers?
S: B and C are closest. → That's funny because in B both the numbers rounded down. → And in C, both numbers rounded up. → So, it's different for subtraction than for addition. If we round them the same way, the difference is closer. → With addition the answer was closer when one number rounded down and one rounded up.

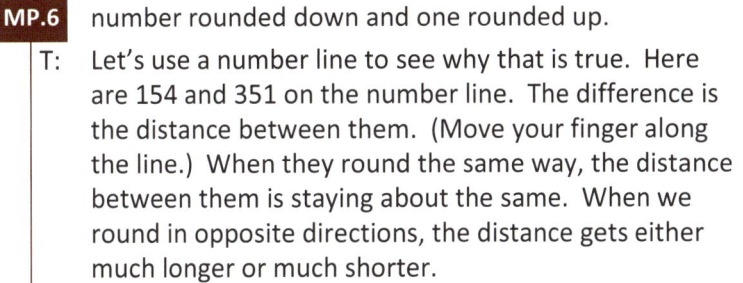

MP.6

T: Let's use a number line to see why that is true. Here are 154 and 351 on the number line. The difference is the distance between them. (Move your finger along the line.) When they round the same way, the distance between them is staying about the same. When we round in opposite directions, the distance gets either much longer or much shorter.
S: That reminds me of my mental math strategy for subtraction. Let's say I'm subtracting 198 from 532. I can just add 2 to both numbers so I have 534 – 200, and the answer is exactly right, 334.
T: Yes. When we add the same number to both the total and the part when we subtract, the difference is still exactly the same.
T: Turn and talk to your partner about what the number line is showing us about estimates when subtracting. Think about what happens on the number line when we add estimates, too. (Allow time for discussion.)
T: Why do we want our estimated differences to be about right? (Allow time for discussion.)
T: Would all four of these estimates help you to check if your exact answer is reasonable?
S: If we used B or C, our exact answer is really close. A is way too small. D is way too big.
T: Just like when we add, we need a good estimate to see if our actual difference is reasonable.

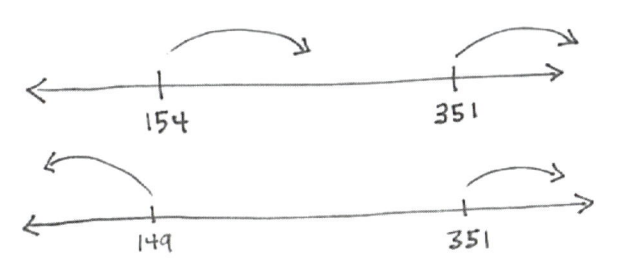

> **NOTES ON MULTIPLE MEANS OF ENGAGEMENT:**
>
> Upon evaluating the usefulness of rounding to the nearest ten or hundred, invite students to propose a better method of rounding to check the reasonableness of answers. For example, students may conclude that rounding to the nearest fifty or nearest twenty-five proves more useful.

> **NOTES ON MULTIPLE MEANS OF ENGAGEMENT:**
>
> This discussion challenges students to identify rules and principles for making best estimates, an activity ideal for students working above grade level. Give students an opportunity to experiment with making best estimates. Guide students to gather their reflections and conclusions in a graphic organizer, such as a flow chart or table, or perhaps as a song, rap, or poem.

Lesson 20: Estimate differences by rounding and apply to solve measurement word problems.

A STORY OF UNITS — Lesson 20 3•2

Problem 3: Round to estimate the difference of 496 − 209. Analyze how rounding to the nearest hundred is nearly the same as rounding to the nearest ten when both addends are close to 1 hundred.

T: (Write the problem above on the board.) With your partner, think about how to round to get the most precise estimate.

Have students analyze the rounded total and part before calculating to determine which is best for a precise answer. Then, have students calculate the estimated difference by rounding to different units. Have them compare estimated answers and then compare with the actual answer.

$$500 - 200 = 300$$
$$500 - 210 = 290$$
$$496 - 200 = 296$$

"This time rounding to the nearest hundred or ten or rounding the part were all pretty close."

$$496 - 209 = 287$$

Application Problem (15 minutes)

Millie's fish tank holds 403 liters of water. She empties out 185 liters of water to clean the tank. How many liters of water are left in the tank?

a. Estimate how many liters are left in the tank by rounding.
b. Estimate how many liters are left in the tank by rounding in a different way.
c. How many liters of water are actually left in the tank?
d. Is your answer reasonable? Which estimate was closer to the exact answer?

T: To solve Part (a), first determine how you are going to round your numbers.
S: (Work and possibly share with a partner.)
T: (Invite a few students to share with the class how they rounded.)

Note: This problem reviews estimating differences to solve word problems, which students learned in today's Concept Development.

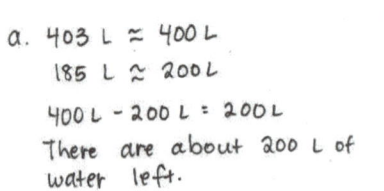

NOTE ON TIMING:

The Problem Set in this lesson is allotted 10 minutes. It directly follows the Application Problem, and so the 10 minutes are included within the 15 minutes allotted for the Application Problem.

a. 403 L ≈ 400 L
   185 L ≈ 200 L
   400 L − 200 L = 200 L
   There are about 200 L of water left.

b. 403 L ≈ 400 L
   185 L ≈ 190 L
   400 L − 190 L = 210 L
   200 / 200
   200 − 190 = 10
   200 + 10 = 210
   There are about 210 L of water left.

c. $\overset{3\;\cancel{10}13}{403}$ L
   − 185 L
   ─────
   218 L
   There are exactly 218 L of water left.

d. Yes, my answer is reasonable because it is close to both of my estimates. Rounding to the nearest 10 L was closer to the actual answer than rounding to the nearest 100 L.

Lesson 20: Estimate differences by rounding and apply to solve measurement word problems.

Problem Set (10 minutes)

Students should do their personal best to complete the Problem Set within the allotted 10 minutes. For some classes, it may be appropriate to modify the assignment by specifying which problems they work on first. Some problems do not specify a method for solving. Students should solve these problems using the RDW approach used for Application Problems.

Student Debrief (10 minutes)

Lesson Objective: Estimate differences by rounding and apply to solve measurement word problems.

The Student Debrief is intended to invite reflection and active processing of the total lesson experience.

Invite students to review their solutions for the Problem Set. They should check work by comparing answers with a partner before going over answers as a class. Look for misconceptions or misunderstandings that can be addressed in the Debrief. Guide students in a conversation to debrief the Problem Set and process the lesson.

Any combination of the questions below may be used to lead the discussion.

- Share your observations from Problem 1(b). What did you find out? How is this different than rounding when you add?
- With your partner, compare your methods of estimation in Problems 2(a) and 3(a). Which was a more precise estimate? If you rounded in the same way, think of another way to estimate. Compare both estimates to the actual answer, and explain why one is more precise than the other.
- When do you need to round so that mental math is easy and fast? When do you need to round more precisely?

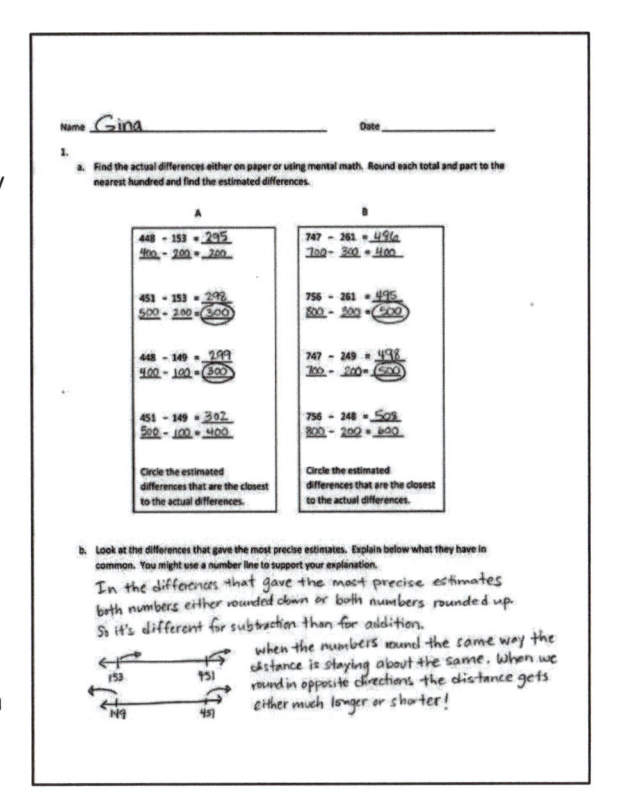

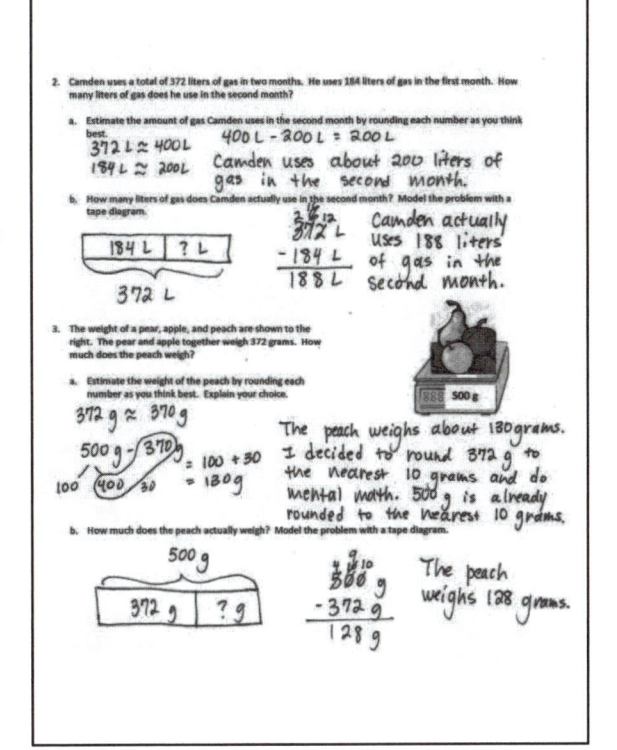

Exit Ticket (3 minutes)

After the Student Debrief, instruct students to complete the Exit Ticket. A review of their work will help with assessing students' understanding of the concepts that were presented in today's lesson and planning more effectively for future lessons. The questions may be read aloud to the students.

A

Round to the Nearest Hundred

Number Correct: _____

1.	201 ≈		23.	350 ≈	
2.	301 ≈		24.	1,350 ≈	
3.	401 ≈		25.	450 ≈	
4.	801 ≈		26.	5,450 ≈	
5.	1,801 ≈		27.	850 ≈	
6.	2,801 ≈		28.	6,850 ≈	
7.	3,801 ≈		29.	649 ≈	
8.	7,801 ≈		30.	651 ≈	
9.	290 ≈		31.	691 ≈	
10.	390 ≈		32.	791 ≈	
11.	490 ≈		33.	891 ≈	
12.	890 ≈		34.	991 ≈	
13.	1,890 ≈		35.	995 ≈	
14.	2,890 ≈		36.	998 ≈	
15.	3,890 ≈		37.	9,998 ≈	
16.	7,890 ≈		38.	7,049 ≈	
17.	512 ≈		39.	4,051 ≈	
18.	2,512 ≈		40.	8,350 ≈	
19.	423 ≈		41.	3,572 ≈	
20.	3,423 ≈		42.	9,754 ≈	
21.	677 ≈		43.	2,915 ≈	
22.	4,677 ≈		44.	9,996 ≈	

Lesson 20: Estimate differences by rounding and apply to solve measurement word problems.

B

Round to the Nearest Hundred

1.	101 ≈	
2.	201 ≈	
3.	301 ≈	
4.	701 ≈	
5.	1,701 ≈	
6.	2,701 ≈	
7.	3,701 ≈	
8.	8,701 ≈	
9.	190 ≈	
10.	290 ≈	
11.	390 ≈	
12.	790 ≈	
13.	1,790 ≈	
14.	2,790 ≈	
15.	3,790 ≈	
16.	8,790 ≈	
17.	412 ≈	
18.	2,412 ≈	
19.	523 ≈	
20.	3,523 ≈	
21.	877 ≈	
22.	4,877 ≈	

23.	250 ≈	
24.	1,250 ≈	
25.	350 ≈	
26.	5,350 ≈	
27.	750 ≈	
28.	6,750 ≈	
29.	649 ≈	
30.	652 ≈	
31.	692 ≈	
32.	792 ≈	
33.	892 ≈	
34.	992 ≈	
35.	996 ≈	
36.	999 ≈	
37.	9,999 ≈	
38.	4,049 ≈	
39.	2,051 ≈	
40.	7,350 ≈	
41.	4,572 ≈	
42.	8,754 ≈	
43.	3,915 ≈	
44.	9,997 ≈	

Number Correct: _____

Improvement: _____

Lesson 20: Estimate differences by rounding and apply to solve measurement word problems.

Name _____ Date _____

1. a. Find the actual differences either on paper or using mental math. Round each total and part to the nearest hundred and find the estimated differences.

A

448 − 153 = _____
____ − ____ = _____

451 − 153 = _____
____ − ____ = _____

448 − 149 = _____
____ − ____ = _____

451 − 149 = _____
____ − ____ = _____

Circle the estimated differences that are the closest to the actual differences.

B

747 − 261 = _____
____ − ____ = _____

756 − 261 = _____
____ − ____ = _____

747 − 249 = _____
____ − ____ = _____

756 − 248 = _____
____ − ____ = _____

Circle the estimated differences that are the closest to the actual differences.

b. Look at the differences that gave the most precise estimates. Explain below what they have in common. You might use a number line to support your explanation.

2. Camden uses a total of 372 liters of gas in two months. He uses 184 liters of gas in the first month. How many liters of gas does he use in the second month?

   a. Estimate the amount of gas Camden uses in the second month by rounding each number as you think best.

   b. How many liters of gas does Camden actually use in the second month? Model the problem with a tape diagram.

3. The weight of a pear, apple, and peach are shown to the right. The pear and apple together weigh 372 grams. How much does the peach weigh?

   a. Estimate the weight of the peach by rounding each number as you think best. Explain your choice.

   b. How much does the peach actually weigh? Model the problem with a tape diagram.

Name _____ Date _____

Kathy buys a total of 416 grams of frozen yogurt for herself and a friend. She buys 1 large cup and 1 small cup.

Large Cup	363 grams
Small Cup	? grams

a. Estimate how many grams are in the small cup of yogurt by rounding.

b. Estimate how many grams are in the small cup of yogurt by rounding in a different way.

c. How many grams are actually in the small cup of yogurt?

d. Is your answer reasonable? Which estimate was closer to the exact weight? Explain why.

Name _____   Date _____

Estimate, and then solve each problem.

1. Melissa and her mom go on a road trip. They drive 87 kilometers before lunch. They drive 59 kilometers after lunch.

   a. Estimate how many more kilometers they drive before lunch than after lunch by rounding to the nearest 10 kilometers.

   b. Precisely how much farther do they drive before lunch than after lunch?

   c. Compare your estimate from (a) to your answer from (b). Is your answer reasonable? Write a sentence to explain your thinking.

2. Amy measures ribbon. She measures a total of 393 centimeters of ribbon and cuts it into two pieces. The first piece is 184 centimeters long. How long is the second piece of ribbon?

   a. Estimate the length of the second piece of ribbon by rounding in two different ways.

   b. Precisely how long is the second piece of ribbon? Explain why one estimate was closer.

3. The weight of a chicken leg, steak, and ham are shown to the right. The chicken and the steak together weigh 341 grams. How much does the ham weigh?

   989 grams

   a. Estimate the weight of the ham by rounding.

   b. How much does the ham actually weigh?

4. Kate uses 506 liters of water each week to water plants. She uses 252 liters to water the plants in the greenhouse. How much water does she use for the other plants?

   a. Estimate how much water Kate uses for the other plants by rounding.

   b. Estimate how much water Kate uses for the other plants by rounding a different way.

   c. How much water does Kate actually use for the other plants? Which estimate was closer? Explain why.

A STORY OF UNITS   Lesson 21   3•2

Lesson 21

Objective: Estimate sums and differences of measurements by rounding, and then solve mixed word problems.

Suggested Lesson Structure

- ■ Fluency Practice   (13 minutes)
- ■ Application Problem   (5 minutes)
- ■ Concept Development   (32 minutes)
- ■ Student Debrief   (10 minutes)
- **Total Time**   **(60 minutes)**

Fluency Practice (13 minutes)

- Group Counting   **3.OA.1**   (4 minutes)
- Use Algorithms with Different Units   **3.MD.2**   (5 minutes)
- Estimate and Subtract   **3.NBT.2**   (4 minutes)

Group Counting (4 minutes)

Note: Group counting reviews interpreting multiplication as repeated addition. It reviews foundational strategies for multiplication from Module 1 and anticipates Module 3.

Direct students to count forward and backward, occasionally changing the direction of the count:

- Threes to 30
- Fours to 40
- Sixes to 60
- Sevens to 70
- Eights to 80
- Nines to 90

As students' fluency with skip-counting improves, help them make a connection to multiplication by tracking the number of groups they count using their fingers.

A STORY OF UNITS — Lesson 21 3•2

Use Algorithms with Different Units (5 minutes)

Materials: (S) Personal white board

Note: This activity reviews addition and subtraction using the standard algorithm.

T: (Write 495 L + 126 L = ___.) On your personal white board, solve using the standard algorithm.

Repeat the process, using the following suggested sequence: 368 cm + 132 cm, 479 cm + 221 cm, 532 cm + 368 cm, 870 L – 39 L, 870 L – 439 L, 807 g – 45 g, and 807 g – 445 g.

Estimate and Subtract (4 minutes)

Materials: (S) Personal white board

Note: This activity reviews rounding to estimate differences from Lesson 20.

T: (Write 71 – 23 ≈ ___.) Say the subtraction sentence.
S: 71 – 23.
T: Say the subtraction sentence, rounding each number to the nearest ten.
S: 70 – 20.
T: (Write 71 – 23 ≈ 70 – 20.) What's 70 – 20?
S: 50.
T: So, 71 – 23 should be close to…?
S: 50.
T: On your boards, answer 71 – 23.
S: (Solve.)

Continue with the following suggested sequence: 47 – 18, 574 – 182, and 704 – 187.

NOTES ON MULTIPLE MEANS OF ACTION AND EXPRESSION:

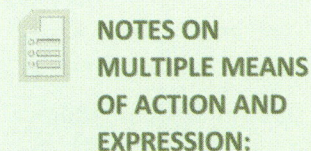

Some learners may be more successful estimating and subtracting if allowed support (without stigma), such as base ten blocks, a place value chart, or a calculator. Maintain high expectations of student achievement, and set realistic personalized goals that they are steadily guided to attain.

Application Problem (5 minutes)

Gloria fills water balloons with 238 mL of water. How many milliliters of water are in two water balloons? Estimate to the nearest 10 mL and 100 mL. Which gives a closer estimate?

Partner 1
238 mL ≈ 240 mL
240 + 240 = 480 mL

? Total mL
[238 mL | 238 mL]

Partner 2
238 mL ≈ 200 mL
200 mL + 200 mL = 400 mL

  238
+ 238

  476

There are 476 mL in 2 balloons.

Rounding to the nearest 10 gives a closer estimate than rounding to the nearest 100.

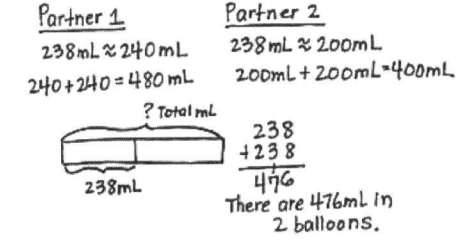

NOTES ON APPLICATION PROBLEM:

Have students complete the problem in partners so that Partner 1 rounds to the nearest ten and Partner 2 rounds to the nearest hundred—it's a time-efficient way of having both estimates to compare with the actual answer.

Note: This problem reviews Lesson 17 by having students round to estimate sums and then calculate the actual answer. It reviews addition because this lesson includes mixed practice with addition and subtraction.

A STORY OF UNITS   Lesson 21   3•2

Concept Development (32 minutes)

Materials: See complete description below.

Problems 1–3 of the Problem Set:

Each group has the premeasured items and measurement tools listed below. Students work together to measure weight, length, and capacity.

Next, they round to estimate sums and differences, and then use the standard algorithm to solve. Determine whether students work in pairs, groups, or individually based on ability. Students should use their estimates to assess the reasonableness of actual answers.

Student Directions: Follow the Problem Set directions to complete Problems 1–3 with your group. Once you have finished those problems, complete Problem 4 on your own.

Materials Description (per group):

Problem 1: 1 digital scale, 1 bag of rice pre-measured at 58 grams, 1 bag of beans pre-measured at 91 grams

Problem 2: 1 meter stick; 3 pieces of yarn labeled A, B, and C (Yarn A pre-measured at 64 cm, Yarn B pre-measured at 88 cm, Yarn C pre-measured at 38 cm)

Problem 3: 1 400 mL beaker, Container D with liquid pre-measured at 212 mL, Container E with liquid pre-measured at 238 mL, Container F with liquid pre-measured at 195 mL

Problem 4: No additional materials

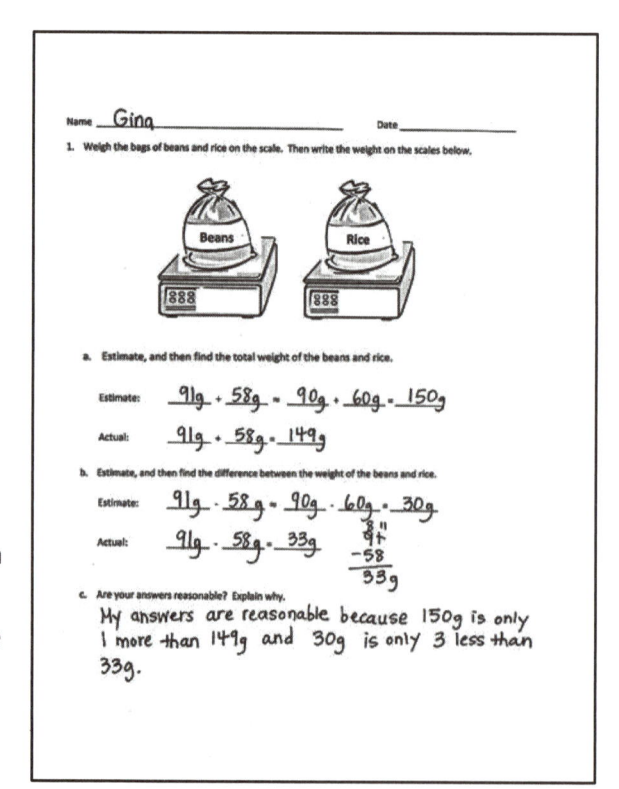

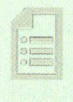

> **NOTES ON MULTIPLE MEANS OF ACTION AND EXPRESSIONS:**
>
> English language learners and others benefit from a demonstration of the procedure, as well as a review of behavior norms. For example, how will turns be recognized? What can be said to request the use of a tool? What is each tool called?
>
> Working in pairs may be to the advantage of English language learners because it provides an opportunity to speak about math in English.

Student Debrief (10 minutes)

Lesson Objective: Estimate sums and differences of measurements by rounding, and then solve mixed word problems.

The Student Debrief is intended to invite reflection and active processing of the total lesson experience.

Invite students to review their solutions for the Problem Set. They should check work by comparing answers with a partner before going over answers as a class.

Look for misconceptions or misunderstandings that can be addressed in the Debrief. Guide students in a conversation to debrief the Problem Set and process the lesson.

Lesson 21: Estimate sums and differences of measurements by rounding, and then solve mixed word problems.

A STORY OF UNITS   Lesson 21   3•2

Any combination of the questions below may be used to lead the discussion.

- How can you use measurement as a tool for checking whether or not your answers are reasonable?
- How did you use mental math in today's lesson?
- How did the Application Problem prepare you for today's Problem Set?
- How did the Fluency Practice relate to your work today?

Exit Ticket (3 minutes)

After the Student Debrief, instruct students to complete the Exit Ticket. A review of their work will help with assessing students' understanding of the concepts that were presented in today's lesson and planning more effectively for future lessons. The questions may be read aloud to the students.

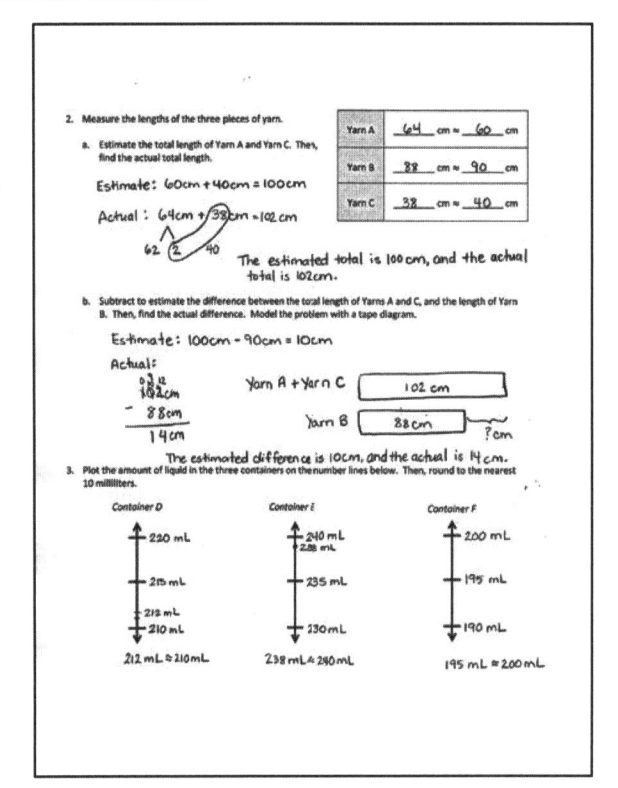

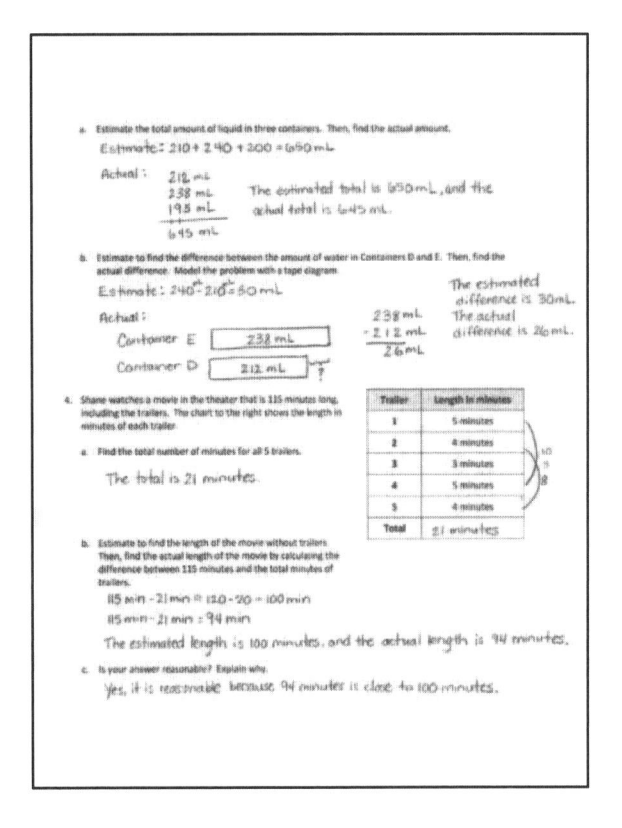

Lesson 21: Estimate sums and differences of measurements by rounding, and then solve mixed word problems.

A STORY OF UNITS

Lesson 21 Problem Set 3•2

Name _____ Date _____

1. Weigh the bags of beans and rice on the scale. Then, write the weight on the scales below.

   a. Estimate, and then find the total weight of the beans and rice.

   Estimate: _____ + _____ ≈ _____ + _____ = _____

   Actual: _____ + _____ = _____

   b. Estimate, and then find the difference between the weight of the beans and rice.

   Estimate: _____ − _____ ≈ _____ − _____ = _____

   Actual: _____ − _____ = _____

   c. Are your answers reasonable? Explain why.

Lesson 21: Estimate sums and differences of measurements by rounding, and then solve mixed word problems.

2. Measure the lengths of the three pieces of yarn.

   a. Estimate the total length of Yarn A and Yarn C. Then, find the actual total length.

Yarn A	_____ cm ≈ _____ cm
Yarn B	_____ cm ≈ _____ cm
Yarn C	_____ cm ≈ _____ cm

   b. Subtract to estimate the difference between the total length of Yarns A and C, and the length of Yarn B. Then, find the actual difference. Model the problem with a tape diagram.

3. Plot the amount of liquid in the three containers on the number lines below. Then, round to the nearest 10 milliliters.

   Container D   Container E   Container F

A STORY OF UNITS

Lesson 21 Problem Set 3•2

a. Estimate the total amount of liquid in three containers. Then, find the actual amount.

b. Estimate to find the difference between the amount of water in Containers D and E. Then, find the actual difference. Model the problem with a tape diagram.

4. Shane watches a movie in the theater that is 115 minutes long, including the trailers. The chart to the right shows the length in minutes of each trailer.

   a. Find the total number of minutes for all 5 trailers.

   b. Estimate to find the length of the movie without trailers. Then, find the actual length of the movie by calculating the difference between 115 minutes and the total minutes of trailers.

Trailer	Length in minutes
1	5 minutes
2	4 minutes
3	3 minutes
4	5 minutes
5	4 minutes
Total	

   c. Is your answer reasonable? Explain why.

A STORY OF UNITS

Lesson 21 Exit Ticket 3•2

Name _____   Date _____

Rogelio drinks water at every meal. At breakfast, he drinks 237 milliliters. At lunch, he drinks 300 milliliters. At dinner, he drinks 177 milliliters.

a. Estimate the total amount of water Rogelio drinks. Then, find the actual amount of water he drinks at all three meals.

b. Estimate how much more water Rogelio drinks at lunch than at dinner. Then, find how much more water Rogelio actually drinks at lunch than at dinner.

A STORY OF UNITS   Lesson 21 Homework 3•2

Name _____   Date _____

1. There are 153 milliliters of juice in 1 carton. A three-pack of juice boxes contains a total of 459 milliliters.

    a. Estimate, and then find the actual total amount of juice in 1 carton and in a three-pack of juice boxes.

       153 mL + 459 mL ≈ _____ + _____ = _____

       153 mL + 459 mL = _____

    b. Estimate, and then find the actual difference between the amount in 1 carton and in a three-pack of juice boxes.

       459 mL − 153 mL ≈ _____ − _____ = _____

       459 mL − 153 mL = _____

    c. Are your answers reasonable? Why?

2. Mr. Williams owns a gas station. He sells 367 liters of gas in the morning, 300 liters of gas in the afternoon, and 219 liters of gas in the evening.

    a. Estimate, and then find the actual total amount of gas he sells in one day.

    b. Estimate, and then find the actual difference between the amount of gas Mr. Williams sells in the morning and the amount he sells in the evening.

Lesson 21: Estimate sums and differences of measurements by rounding, and then solve mixed word problems.

3. The Blue Team runs a relay. The chart shows the time, in minutes, that each team member spends running.

   a. How many minutes does it take the Blue Team to run the relay?

Blue Team	Time in Minutes
Jen	5 minutes
Kristin	7 minutes
Lester	6 minutes
Evy	8 minutes
Total	

   b. It takes the Red Team 37 minutes to run the relay. Estimate, and then find the actual difference in time between the two teams.

4. The lengths of three banners are shown to the right.

   a. Estimate, and then find the actual total length of Banner A and Banner C.

Banner A	437 cm
Banner B	457 cm
Banner C	332 cm

   b. Estimate, and then find the actual difference in length between Banner B and the combined length of Banner A and Banner C. Model the problem with a tape diagram.

A STORY OF UNITS   End-of-Module Assessment Task 3•2

Name _____   Date _____

1. Paul is moving to Australia. The total weight of his 4 suitcases is shown on the scale to the right. On a number line, round the total weight to the nearest 100 kilograms.

2. Paul buys snacks for his flight. He compares cashews to yogurt raisins. The cashews weigh 205 grams, and the yogurt raisins weigh 186 grams. What is the difference between the weight of the cashews and yogurt raisins?

3. The clock to the right shows what time it is now.

   Time Right Now

   a. Estimate the time to the nearest 10 minutes.

   b. The clock to the right show Paul's departure time. Estimate the time to the nearest 10 minutes.

   Departure Time

   c. Use your answers from Parts (a) and (b) to estimate how long Paul has before his flight leaves.

4. A large airplane uses about 256 liters of fuel every minute.

   a. Round to the nearest ten liters to estimate how many liters of fuel get used every minute.

   b. Use your estimate to find about how many liters of fuel are used every 2 minutes.

   c. Calculate precisely how many liters of fuel are used every 2 minutes.

   d. Draw a tape diagram to find the difference between your estimate and the precise calculation.

5. Baggage handlers lift heavy luggage into the plane. The weight of one bag is shown on the scale to the right.

   a. One baggage handler lifts 3 bags of the same weight. Round to estimate the total weight he lifts. Then, calculate exactly.

   b. Another baggage handler lifts luggage that weighs a total of 200 kilograms. Write and solve an equation to show how much more weight he lifts than the first handler in Part (a).

   c. The baggage handlers load luggage for 18 minutes. If they start at 10:25 p.m., what time do they finish?

   d. One baggage handler drinks the amount of water shown below every day at work. How many liters of water does he drink during all 7 days of the week?

6. Complete as many problems as you can in 100 seconds. The teacher will time you and tell you when to stop.

3 x 1 = _____   2 ÷ 1 = _____   _____ = 20 ÷ 10   2 x 2 = _____   5 x _____ = 10

_____ x 2 = 4   10 ÷ 5 = _____   10 x _____ = 30   _____ = 2 x 3   _____ = 12 ÷ 4

4 x 3 = _____   15 ÷ 5 = _____   _____ x 4 = 16   _____ = 40 ÷ 10   2 x 4 = _____

3 x 4 = _____   4 x _____ = 12   20 ÷ 4 = _____   _____ = 10 x 5   _____ x 5 = 25

4 x _____ = 20   _____ = 10 ÷ 2   _____ x 3 = 18   10 x 6 = _____   30 ÷ 5 = _____

3 x 6 = _____   _____ = 24 ÷ 4   5 x _____ = 35   _____ = 10 x 7   14 ÷ 2 = _____

2 x 7 = _____   _____ x 4 = 28   _____ = 40 ÷ 5   10 x _____ = 80   _____ = 3 x 8

24 ÷ 3 = _____   80 ÷ 10 = _____   36 ÷ 4 = _____   5 x 9 = _____   2 x _____ = 18

| A STORY OF UNITS | End-of-Module Assessment Task | 3•2 |

End-of-Module Assessment Task
Standards Addressed — Topics A–F

Use place value understanding and properties of operations to perform multi-digit arithmetic. (A range of algorithms may be used.)

- **3.NBT.1** Use place value understanding to round whole numbers to the nearest 10 or 100.
- **3.NBT.2** Fluently add and subtract within 1000 using strategies and algorithms based on place value, properties of operations, and/or the relationship between addition and subtraction.

Solve problems involving measurement and estimation of intervals of time, liquid volumes, and masses of objects.

- **3.MD.1** Tell and write time to the nearest minute and measure time intervals in minutes. Solve word problems involving addition and subtraction of time intervals in minutes, e.g., by representing the problem on a number line diagram.
- **3.MD.2** Measure and estimate liquid volumes and masses of objects using standard units of grams (g), kilograms (kg), and liters (l). (Excludes compound units such as cm^3 and finding the geometric volume of a container.) Add, subtract, multiply, or divide to solve one-step word problems involving masses or volumes that are given in the same units, e.g., by using drawings (such as a beaker with a measurement scale) to represent the problem. (Excludes multiplicative comparison problems, i.e., problems involving notions of "times as many"; see CCLS Glossary, Table 2.)

Multiply and divide within 100.

- **3.OA.7** Fluently multiply and divide within 100, using strategies such as the relationship between multiplication and division (e.g., knowing that 8 × 5 = 40, one knows 40 ÷ 5 = 8) or properties of operations. By the end of Grade 3, know from memory all products of two one-digit numbers.

Evaluating Student Learning Outcomes

A Progression Toward Mastery is provided to describe steps that illuminate the gradually increasing understandings that students develop *on their way to proficiency.* In this chart, this progress is presented from left (Step 1) to right (Step 4) for Problems 1–5. The learning goal for students is to achieve Step 4 mastery. These steps are meant to help teachers and students identify and celebrate what students CAN do now and what they need to work on next. Problem 6 is scored differently since it is a timed assessment of fluency. Students complete as many problems as they can in 100 seconds. Although this page of the assessment contains 40 questions, answering 30 correct within the time limit is considered passing.

End-of-Module Assessment Task 3•2

A Progression Toward Mastery

Assessment Task Item	STEP 1 Little evidence of reasoning without a correct answer. (1 Point)	STEP 2 Evidence of some reasoning without a correct answer. (2 Points)	STEP 3 Evidence of some reasoning with a correct answer or evidence of solid reasoning with an incorrect answer. (3 Points)	STEP 4 Evidence of solid reasoning with a correct answer. (4 Points)
1 3.NBT.1 3.MD.2	Student is unable to answer the question correctly. The attempt shows the student may not understand the meaning of the question.	Student attempts to answer the question. Mistakes may include those listed in the box to the right and/or misreading the scale but correctly rounding based on error.	Same criteria as for a 4 but may omit the unit (kg) in one or more parts of the answer.	Student answers the question correctly: ▪ Accurately reads the scale as 127 kg. ▪ Rounds on a number line to estimate 100 kg.
2 3.NBT.2	Student is unable to answer the question correctly. The attempt shows student may not understand the meaning of the question.	Student attempts to answer the question. Mistakes may include those listed in the box to the right and/or decomposing the numbers incorrectly.	Student may or may not answer question correctly. Mistakes may include decomposing the numbers correctly but making a calculation error when subtracting.	Student correctly writes and solves 205 g – 186 g = 19 g.
3 3.NBT.1 3.NBT.2 3.MD.1	Student is unable to answer questions correctly. The attempt shows the student may not understand the meaning of the questions.	Student attempts to answer the questions. Mistakes may include those listed in the box to the right and/or inaccurately reading one or both of the clocks.	Student answers at least one question correctly. Mistakes may include a rounding error in either Part (a) or Part (b) affecting Part (c) or a correctly solved problem based on a wrong answer.	Student answers every question correctly: a. Rounds 10:19 to 10:20. b. Rounds 10:53 to 10:50. c. Estimates about 30 minutes before the plane leaves.

A Progression Toward Mastery					
4 3.NBT.1 3.NBT.2		Student is unable to answer any of the questions correctly. The attempt shows the student may not understand the meaning of the questions.	Student attempts to answer the questions. Mistakes may include those listed in the box to the right, and/or: ▪ Either failing to round or calculate exactly in Parts (a–d). ▪ Omitting the units in any part. ▪ Incorrectly drawing or labelling a tape diagram.	Student may or may not answer questions correctly. Mistakes may include an arithmetic error in Part (c) affecting Part (d) or a tape diagram drawn and labeled correctly based on a wrong answer.	Student answers every question correctly: a. Rounds to estimate 260 liters in Part (a). b. Estimates 520 liters in Part (b). c. Precisely calculates 512 liters in Part (c). d. Draws and labels a tape diagram to show 8 liters as the difference in Part (d).
5 3.NBT.1 3.NBT.2 3.MD.1 3.MD.2		Student is unable to answer any question correctly. The attempt shows the student may not understand the meaning of the questions.	Student attempts to answer the questions. Mistakes may include those listed in the box to the right, and/or: ▪ Conceptual rather than calculation error in Parts (a), (b), or (d). ▪ Either failing to round or calculate exactly in Part (a). ▪ Omitting the units in any part.	Student may or may not answer questions correctly. Mistakes may include those listed below: ▪ Arithmetic error in Part (a) affecting Part (b) but solved correctly based on a wrong answer. ▪ Failing to write an equation in Part (b).	Student answers every question correctly: a. 65 kg + 65 kg + 65 kg = 195 kg, and 70 kg + 70 kg + 70 kg = 210 kg in Part (a). b. Writes and solves 200 kg – 195 kg = 5 kg in Part (b). c. Calculates end time of 10:43 p.m. in Part (c). d. May use multiplication or addition to answer 28 liters in Part (d).
6 3.OA.7		Use the attached sample work to correct students' answers on the fluency page of the assessment. Students who answer 30 or more questions correctly within the allotted time pass this portion of the assessment. They are ready to move on to the more complicated fluency page given with the Module 3 End-of-Module Assessment. For students who do not pass, you may choose to re-administer this fluency page with each subsequent End-of-Module assessment until they are successful. Analyze the mistakes students make on this assessment to further guide your fluency instruction. Possible questions to ask as you analyze are given below: ▪ Did this student struggle with multiplication, division, or both? ▪ Did this student struggle with a particular factor? ▪ Did the student consistently miss problems with the unknown in a particular position?			

Name __Gina__   Date _____

1. Paul is moving to Australia. The total weight of his 4 suitcases is shown on the scale to the right. On a number line, round the total weight to the nearest 100 kilograms.

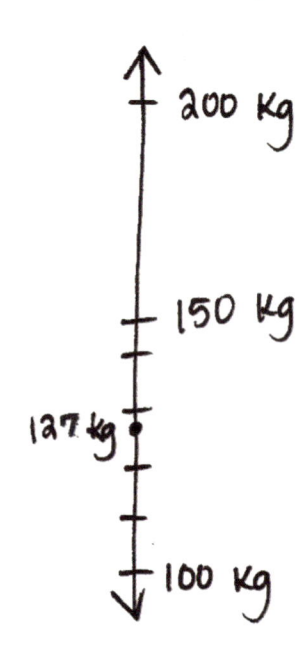

```
↑
┼ 200 kg
│
│
┤ 150 kg
│
│
127 kg •
│
┼ 100 kg
↓
```

Rounded to the nearest 100 kg, his suitcases weighs 100 kg.

[scale reads 127 kg]

2. Paul buys snacks for his flight. He compares cashews with yogurt raisins. The cashews weigh 205 grams, and the yogurt raisins weigh 186 grams. What is the difference between the weight of the cashews and yogurt raisins?

$$\begin{array}{r} \overset{1\;9\;15}{2\cancel{0}\cancel{5}} \text{ g} \\ -1\;8\;6 \text{ g} \\ \hline 0\;1\;9 \text{ g} \end{array}$$

The difference in weight is 19 grams.

3. The clock to the right shows what time it is now.
   a. Estimate the time to the nearest 10 minutes.

   Time right now:

   10:19

   ↑
   ├ 10:20
   10:19 ↗ •
   │
   ├ 10:15
   │
   ├ 10:10
   ↓

   It is 10:20 to the nearest 10 minutes.

   b. The clock to the right show Paul's departure time. Estimate the time to the nearest 10 minutes.

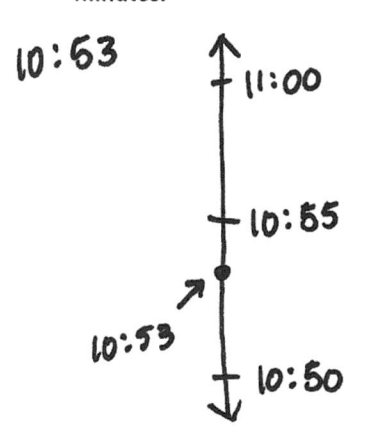

   Departure time:

   10:53

   ↑
   ├ 11:00
   │
   ├ 10:55
   10:53 ↗ •
   ├ 10:50
   ↓

   His departure time is 10:50 to the nearest 10 minutes.

   c. Use your answers from Parts (a) and (b) to estimate how long Paul has before his flight leaves.

   50 minutes - 20 minutes = 30 minutes

   Paul has about 30 minutes before his flight leaves.

A STORY OF UNITS

End-of-Module Assessment Task 3•2

4. A large airplane uses about 256 liters of fuel every minute.

   a. Round to the nearest ten liters to estimate how many liters of fuel get used every minute.

   ↑ 260 L
   256 L •
   ↓ 250 L

   About 260 L of fuel are used every minute.

   b. Use your estimate to find about how many liters of fuel are used every 2 minutes.

   ```
     260 L
   + 260 L
   ─────
     520 L
   ```

   About 520 L of fuel are used every 2 minutes.

   c. Calculate precisely how many liters of fuel are used every 2 minutes.

   ```
     256 L
   + 256 L
   ─────
     512 L
   ```

   Exactly 512 L of fuel are used in 2 minutes.

   d. Draw a tape diagram to find the difference between your estimate and precise calculation.

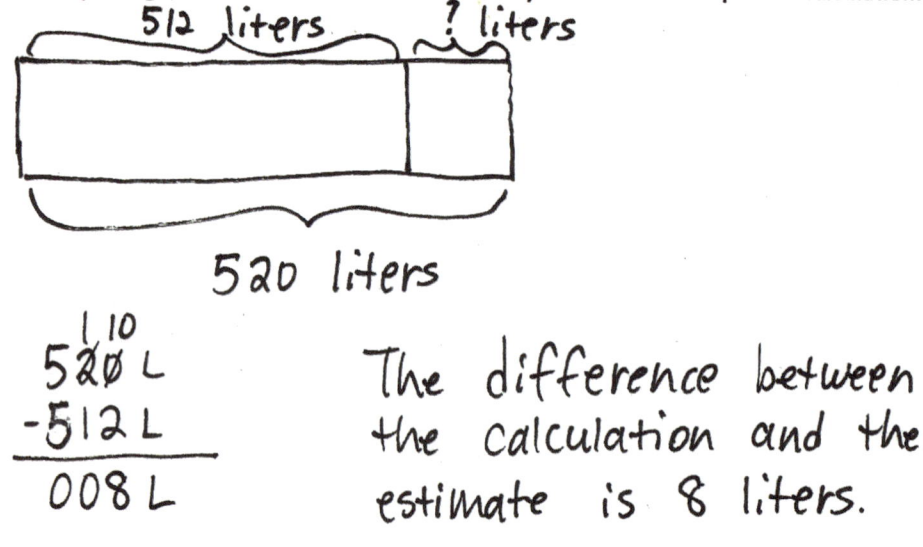

   512 liters | ? liters

   520 liters

   ```
     5̸2̸0 L   (1, 10)
   - 512 L
   ─────
     008 L
   ```

   The difference between the calculation and the estimate is 8 liters.

A STORY OF UNITS   End-of-Module Assessment Task   3•2

5. Baggage handlers lift heavy luggage into the plane. The weight of one bag is shown on the scale to the right.

   a. One baggage handler lifts 3 bags of the same weight. Round to estimate the total weight he lifts. Then, calculate exactly.

   65 kg is about 70 kg.

   $\begin{array}{r} 70 \\ 70 \\ +70 \\ \hline 210 \end{array}$ 14 ⟨ 70, 21 ⟨ 70

   He lifts about 210 kg total.

   $\begin{array}{r} 65 \\ 65 \\ +65 \\ \hline 195 \end{array}$ 12 ⟨ 65, 18 ⟨ 65

   He lifts exactly 195 kg.

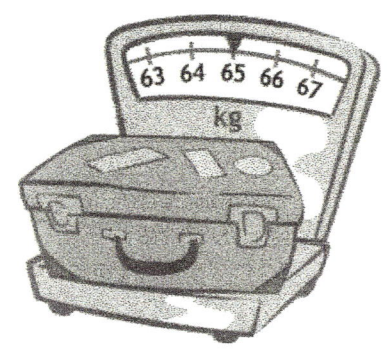

   b. Another baggage handler lifts luggage that weighs a total of 200 kilograms. Write and solve an equation to show how much more weight he lifts than the first handler in Part (a).

   $\begin{array}{r} \cancel{200} \\ -195 \\ \hline 005 \end{array}$

   He lifts 5 kg more than the first handler.

   c. The baggage handlers load luggage for 18 minutes. If they start at 10:25 p.m., what time do they finish?

   $\begin{array}{r} 25 \\ +18 \\ \hline 43 \end{array}$

   They finish at 10:43 p.m.

   d. One baggage handler drinks the amount of water shown below every day at work. How many liters of water does he drink during all 7 days of the week?

   7 × 4 L = 28 L

   He drinks 28 L of water in 7 days.

   4, 8, 12, 16, 20, 24, 28

6. Complete as many problems as you can in 100 seconds. The teacher will time you and tell you when to stop.

3 x 1 = 3   2 ÷ 1 = 2   2 = 20 ÷ 10   2 x 2 = 4   5 x 2 = 10

2 x 2 = 4   10 ÷ 5 = 2   10 x 3 = 30   6 = 2 x 3   3 = 12 ÷ 4

4 x 3 = 12   15 ÷ 5 = 3   4 x 4 = 16   4 = 40 ÷ 10   2 x 4 = 8

3 x 4 = 12   4 x 3 = 12   20 ÷ 4 = 5   50 = 10 x 5   5 x 5 = 25

4 x 5 = 20   5 = 10 ÷ 2   6 x 3 = 18   10 x 6 = 60   30 ÷ 5 = 6

3 x 6 = 18   6 = 24 ÷ 4   5 x 7 = 35   70 = 10 x 7   14 ÷ 2 = 7

2 x 7 = 14   7 x 4 = 28   8 = 40 ÷ 5   10 x 8 = 80   24 = 3 x 8

24 ÷ 3 = 8   80 ÷ 10 = 8   36 ÷ 4 = 9   5 x 9 = 45   2 x 9 = 18

Answer Key

Eureka Math
Grade 3
Module 2

Special thanks go to the Gordan A. Cain Center and to the Department of Mathematics at Louisiana State University for their support in the development of *Eureka Math*.

Published by Great Minds

Copyright © 2015 Great Minds. All rights reserved. No part of this work may be reproduced or used in any form or by any means — graphic, electronic, or mechanical, including photocopying or information storage and retrieval systems — without written permission from the copyright holder. "Great Minds" and "Eureka Math" are registered trademarks of Great Minds.

Printed in the U.S.A.
This book may be purchased from the publisher at eureka-math.org
10 9 8 7 6 5 4 3 2 1

A STORY OF UNITS

Mathematics Curriculum

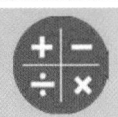

GRADE 3 • MODULE 2

Answer Key
GRADE 3 • MODULE 2

Place Value and Problem Solving with Units of Measure

Lesson 1

Problem Set

1. Times will vary.
2. Times will vary.
3. Times will vary.
4. Times will vary.
5. Times will vary.
6. Times will vary.

Exit Ticket

a. Jake
b. Riley and Nicholas
c. 3 seconds

Homework

1. a. Dominique
   b. Chester
   c. 5 seconds
2. Activities will vary.
3. First clock—10:15
   Second clock—2:50
   Third clock—11:00
   Fourth clock—7:05

Lesson 2

Problem Set

1. a. First and last tick marks labeled as 7:00 a.m. and 8:00 a.m.
   b. Each interval labeled by fives below the number line up to 8:00 a.m.
   c. Point D plotted and labeled above 7:10 a.m.
   d. Point E plotted and labeled above 7:35 a.m.
   e. Point T plotted and labeled above 7:40 a.m.
   f. Point L plotted and labeled above 7:45 a.m.
   g. Point W plotted and labeled above 7:55 a.m.

2. Every 5 minutes labeled below the number line

   First clock not matched to the number line

   Second clock—5:50 p.m.

   Third clock—5:15 p.m.

   Fourth clock not matched to the number line

   Fifth clock—5:40 p.m.

   Last clock—5:25 p.m.

3. First and last tick marks labeled as 5:00 p.m. and 6:00 p.m.; each interval labeled by fives below the number line up to 6:00 p.m.; 5:45 p.m. located and plotted on the number line

4. Answers will vary.

Exit Ticket

a. 10:10 a.m.
b. 10:20 a.m.
c. 10:50 a.m.
d. 1 hour

Homework

a. First and last tick marks labeled as 4:00 p.m. and 5:00 p.m.
b. Each interval labeled by fives below the number line up to 5:00 p.m.
c. Point W plotted and labeled above 4:05 p.m.
d. Point F plotted and labeled above 4:15 p.m.
e. Point G plotted and labeled above 4:25 p.m.
f. Point B plotted and labeled above 4:50 p.m.
g. Point P plotted and labeled above 4:55 p.m.

Lesson 3

Problem Set

1. The times shown on the clocks are plotted correctly on the number line.
   First clock—7:17 p.m.
   Second clock—7:03 p.m.
   Third clock—7:55 p.m.
   Fourth clock—7:41 p.m.
   Fifth clock—answer provided

2. Hands on the clock drawn to show 6:48 a.m.
3. Hands on the clock drawn to show 8:23 a.m.
4. 5:27
5. a. 3:56
   b. 3:45

Exit Ticket

a. 8:04
b. Hands on the clock drawn to show 8:23 a.m.
c. The first and last tick marks labeled as 8:00 a.m. and 9:00 a.m.; Point A plotted and labeled above 8:04 a.m.; Point B plotted and labeled above 8:23 a.m.

Homework

1. The times shown on the clocks are plotted correctly on the number line.
   First clock—4:34 p.m.
   Second clock—4:01 p.m.
   Third clock—4:16 p.m.
   Fourth clock—4:53 p.m.
   Fifth clock—answer provided

2. Hands on the clock drawn to show 6:07 p.m.
3. Hands on the clock drawn to show 1:32 p.m.
4. a. 2:32
   b. 2:55
   c. Hands on the clock drawn to show 2:55.
   d. First and last tick marks labeled 2:00 p.m. and 3:00 p.m.; Point B plotted and labeled above 2:32 p.m.; Point F plotted and labeled above 2:55 p.m.

Lesson 4

Problem Set

1. 26
2. 2:08
3. 31
4. 4:09
5. 9:52
6. 19 min
7. 11:58 a.m.
8. 1:17 p.m.

Exit Ticket

1. Hands on the first clock are drawn to show 1:34 p.m.
2. Hands on the second clock are drawn to show 1:56 p.m.
3. 22 min

Homework

1. 31
2. 3:22
3. 33
4. 2:11
5. 36 min
6. Times will vary.

Lesson 5

Problem Set

1. 53; problem modeled on number line; 25 + 28 = 53
2. 22 minutes; problem modeled on number line; 34 − 12 = 22
3. 17 minutes; 47 − 30 = 17
4. a. 29 minutes
   b. No; Austin will be 4 minutes late.
5. 11:13

Exit Ticket

36; problem modeled on number line; 19 + 17 = 36

Homework

1. 56; problem modeled on number line; 22 + 34 = 56
2. 9 minutes; problem modeled on number line; 56 − 47 = 9
3. 30 minutes
4. a. 47 minutes
   b. No; Marcus will be 2 minutes late.
5. 27 minutes

Lesson 6

Problem Set

1. Illustrations and descriptions will vary.
2. Illustrations and descriptions will vary.
3. Illustrations and descriptions will vary.
4. Illustrations and descriptions will vary.
5. Answers will vary.

Exit Ticket

100 grams

Homework

1. a. 10
   b. 10
   c. 10
   d. They all need 10 to get to the next unit.
2. Top row, left to right: 3 kilograms; 6 kilograms; 450 grams
   Bottom row, left to right: 907 grams; 11 kilograms; 1 kilogram

Lesson 7

Problem Set

A. Objects and weights will vary.

B. Objects and weights will vary.

C. Objects and weights will vary.

D. Objects and weights will vary.

E.
1. grams
2. kilograms
3. grams
4. kilograms
5. kilograms
6. grams

F. 2 kilograms since 1 bottle of water weighs about 1 kilogram

G. Yes; 10 units of 100 grams equal 1000 grams, which is the same as 1 kilogram.

Exit Ticket

1. 146 g; 12 kg
2.
   a. grams
   b. grams
   c. kilograms
   d. grams
   e. kilograms

Homework

1. Water bottle—1 kilogram

   Paper clip—1 gram

   4 pennies—10 grams

   Apple—100 grams

2. Grams; because 113 kilograms is too heavy for a cell phone

3. 25 kilograms; 9 kilograms; 200 grams

   367 grams; 105 grams

Lesson 8

Problem Set

1. 464; 355
2. a. 78; problem modeled with tape diagram
   b. 8; problem modeled with tape diagram
3. Tape diagram drawn correctly; about 15 kg
4. a. About 3 kg
   b. About 21 kg

Exit Ticket

a. 14 kg
b. 28 kg
c. 3 backpacks

Homework

1. a. C
   b. B
   c. 4
   d. 36 kg
2. 840 g
3. 430 g
4. a. 91 kg
   b. 125 kg
5. a. 7 kg
   b. 5 kg

Lesson 9

Problem Set

a. Predictions will vary.

b. Answers will vary.

c. Illustrations and descriptions will vary.

d. Illustrations and descriptions will vary.

e. Illustrations and descriptions will vary.

f. They both break apart into 1 thousand units. 1 liter is 1000 milliliters, and 1 kilogram is 1000 grams.

g. 1 gram; 1 liter is the same as 1 kilogram, and they break apart the same way into 1 thousand units.

Exit Ticket

1. 10; Morgan will scoop water 10 times.

2. 100 groups; there are 10 groups of 10 milliliters in 100 milliliters, and there are 10 groups of 100 milliliters in 1 liter.

Homework

1. a. Answers will vary.

   b. Answers will vary.

2. 15 mL

3. 708 mL

4. 6 buckets

5. 5 L

Lesson 10

Problem Set

1. Vertical number line on container labeled by hundreds
   a. 500 mL; reasons will vary.
   b. Explanations will vary.
   c. 700 mL
2. 3 L; 6 L; 4 L; 0 L
3. 400 mL; 200 mL; 1000 mL; 700 mL
4. a. Capacity of each barrel plotted and labeled correctly on number line
   b. Barrel C
   c. Barrel D
   d. Barrel B because it is closest to 70 mL OR Barrel A because it has enough capacity to hold 70 L
   e. Number line used to find answer; 28 more liters

Exit Ticket

1. A: 45 L
   B: 57 L
   C: 21 L
2. 24 L

Homework

1. 5 L; 2 L; 6 L; 1 L
2. 11 L
3. 5 L; 2 L; 4 L; 2 L
4. a. Capacity of each gas tank plotted and labeled on number line
   b. Large
   c. Small
   d. Medium
   e. Number line used to find answer; 32 more liters

Lesson 11

Problem Set

1. a. 558 g
   b. 445 g
2. a. 60 g
   b. 142 g
3. a. 191 g
   b. 123 g
   c. 194 g
4. Tape diagram drawn and labeled to represent the problem; 9 turkeys
5. 900 mL of milk
6. 14 L

Exit Ticket

a. 677 mL
b. 140 mL
c. 480 mL

Homework

1. 687
2. 104
3. 54 L
4. 8 beds
5. 35 mL

Lesson 12

Problem Set

1. Measurements and estimates will vary.
2. Measurements and estimates will vary.
3. Measurements and estimates will vary.
4. Measurements and estimates will vary.

Exit Ticket

a. 46 g
b. Rounding modeled on number line
c. 50 g
d. 46 g is more than halfway between 40 g and 50 g on the number line, so 46 g rounds up to 50 g.

Homework

1. Measurements and estimates will vary.
2. 10:30
3. 20
4. 53; 50
5. 58; 60

Lesson 13

Problem Set

1. a. 30
   b. 40; rounding modeled on number line
   c. 60; rounding modeled on number line
   d. 160; rounding modeled on number line
   e. 280; rounding modeled on number line
   f. 410; rounding modeled on number line

2. Number line drawn and labeled to model rounding; 40 g
   Number line drawn and labeled to model rounding; 50 g
   Number line drawn and labeled to model rounding; 140 g

3. a. 48 min
   b. 50 min

Exit Ticket

1. a. 30; rounding modeled on number line
   b. 280; rounding modeled on number line

2. No; 603 is less than halfway between 600 and 610, so 603 rounded to the nearest ten is 600; number line drawn and labeled to model rounding

Homework

1. a. 40
   b. 50; rounding modeled on number line
   c. 70; rounding modeled on number line
   d. 170; rounding modeled on number line
   e. 190; rounding modeled on number line
   f. 190; rounding modeled on number line

2. Number line drawn and labeled to model rounding; 50 g
   Number line is drawn and labeled to model rounding; 670 g

3. 60 g; number line drawn and labeled to model rounding

Lesson 14

Sprint

Side A

1. 5
2. 15
3. 25
4. 75
5. 75
6. 45
7. 45
8. 35
9. 35
10. 65
11. 65
12. 85
13. 95
14. 95
15. 85
16. 55
17. 155
18. 255
19. 755
20. 755
21. 85
22. 185
23. 285
24. 585
25. 585
26. 35
27. 935
28. 65
29. 465
30. 95
31. 895
32. 995
33. 1,005
34. 75
35. 1,075
36. 1,575
37. 485
38. 1,485
39. 1,085
40. 355
41. 1,785
42. 395
43. 1,835
44. 1,105

Side B

1. 15
2. 25
3. 35
4. 65
5. 65
6. 55
7. 55
8. 45
9. 45
10. 75
11. 75
12. 85
13. 95
14. 95
15. 85
16. 65
17. 165
18. 265
19. 565
20. 565
21. 75
22. 175
23. 275
24. 675
25. 675
26. 25
27. 925
28. 55
29. 455
30. 95
31. 895
32. 995
33. 1,005
34. 25
35. 1,025
36. 1,525
37. 385
38. 1,385
39. 1,085
40. 755
41. 1,685
42. 295
43. 1,845
44. 1,215

Lesson 14 Answer Key 3•2

Problem Set

1. a. 100; rounding modeled on number line
   b. 300; rounding modeled on number line
   c. 300; rounding modeled on number line
   d. 1,300; rounding modeled on number line
   e. 1,600; rounding modeled on number line
   f. 1,300; rounding modeled on number line

2. a. 500 stickers
   b. 500 pages
   c. 800 mL
   d. $1,300
   e. 1,800 km

3. 550, 639, 603

4. Both are correct; explanations will vary.

Exit Ticket

1. a. 100; rounding modeled on number line
   b. 1,800; rounding modeled on number line

2. 700 people; vertical number line drawn

Homework

1. a. 200; rounding modeled on number line
   b. 300; rounding modeled on number line
   c. 300; rounding modeled on number line
   d. 1,300; rounding modeled on number line
   e. 1,700; rounding modeled on number line
   f. 1,800; rounding modeled on number line

2. a. 200 cards
   b. 500 people
   c. 400 milliliters
   d. 700 grams
   e. $1,300

3. 368, 420, 449

4. Both are correct; explanations will vary.

Lesson 15

Problem Set

1. a. 51 mL
   b. 71 mL
   c. 171 mL
   d. 89 cm
   e. 592 cm
   f. 627 cm
   g. 92 g
   h. 639 g
   i. 956 g
   j. 3 L 657 mL
   k. 5 kg 876 g

2. 107 g
3. 475 mL + 317 mL = 792 mL; Andrea is correct; explanations will vary.
4. 47 min

Exit Ticket

1. a. 60 cm
   b. 742 m
   c. 584 km

2. a. 41 min
   b. 67 min

Homework

1. a. 82 cm
   b. 95 kg
   c. 591 mL
   d. 375 g
   e. 790 mL
   f. 480 L
2. a. 373
   b. 444 mL
3. 119 students
4. 63 cm
5. Paperback book and bar of soap; 343 g + 117 g = 460 g

Lesson 16

Problem Set

1. a. 120 mL
    b. 420 mL
    c. 820 mL
    d. 150 cm
    e. 600 cm
    f. 900 cm
    g. 835 g
    h. 942 g
    i. 983 g
    j. 4 L 800 mL
    k. 6 kg 851 g

2. Tape diagram drawn and labeled; 1,000 g
3. 144 muffins
4. 741 mL

Exit Ticket

1. a. 107 g
    b. 617 kg
    c. 802 L
2. 104 L

Homework

1. a. 55 m
    b. 85 m
    c. 530 m
    d. 72 mL
    e. 542 mL
    f. 642 mL
    g. 631 kg
    h. 801 kg
    i. 902 kg
    j. 6 L 556 mL
    k. 8 kg 622 g

2. Tape diagram drawn and labeled; 101 minutes
3. 324
4. 802

Lesson 17

Sprint

Side A

1. 20
2. 30
3. 40
4. 80
5. 60
6. 50
7. 40
8. 20
9. 40
10. 30
11. 60
12. 50
13. 80
14. 70
15. 70
16. 60
17. 30
18. 40
19. 50
20. 80
21. 90
22. 20
23. 80
24. 90
25. 100
26. 110
27. 120
28. 150
29. 310
30. 410
31. 520
32. 620
33. 630
34. 640
35. 670
36. 970
37. 980
38. 990
39. 1,000
40. 1,110
41. 1,120
42. 3,230
43. 5,490
44. 7,890

Side B

1. 10
2. 20
3. 30
4. 70
5. 70
6. 60
7. 50
8. 20
9. 30
10. 20
11. 50
12. 40
13. 90
14. 80
15. 80
16. 70
17. 20
18. 30
19. 40
20. 80
21. 90
22. 50
23. 80
24. 90
25. 100
26. 110
27. 120
28. 160
29. 210
30. 310
31. 420
32. 520
33. 530
34. 540
35. 570
36. 970
37. 980
38. 990
39. 1,000
40. 1,110
41. 1,120
42. 2,340
43. 4,580
44. 8,790

A STORY OF UNITS

Lesson 17 Answer Key 3•2

Problem Set

1. a. A: 704; 500, 300, 800
       700; 500, 200, (700)
       697; 400, 200, 600
     B: 517; 400, 200, 600
       504; 400, 100, (500)
       496; 300, 100, 400
     C: 810; 700, 200, 900
       805; 600, 200, (800)
       793; 600, 100, 700
   b. Explanations will vary; both addends are close to the halfway point, so they balance each other out.

2. a. Estimates will vary.
   b. 245 min
   c. Explanations will vary; a different way of rounding is shown and compared.

3. a. Estimates will vary.
   b. 256 kilograms; a tape diagram is drawn and labeled to represent the problem.

Exit Ticket

a. 420 minutes
b. 400 minutes
c. Explanations will vary; both addends are close to the halfway point, so rounding to the nearest 10 minutes and 100 minutes give estimates that are close to each other.

Homework

1. a. 40 kg
   b. 39 kg
   c. 70 min
   d. 61 min
   e. A close estimate can help us see if our actual sum is reasonable.
2. a. Estimates will vary.
   b. Estimates will vary.
   c. 573 min; explanations will vary.

Lesson 18

Problem Set

1. a. 36 mL
    b. 336 mL
    c. 136 mL
    d. 497 cm
    e. 361 cm
    f. 498 cm
    g. 177 g
    h. 73 g
    i. 75 g
    j. 1 km 315 m
    k. 2 kg 31 g

2. 172 g; tape diagram drawn and labeled to model problem
3. a. 95 min
    b. 50 min
4. 34 cm

Exit Ticket

1. a. 235 mL
    b. 304 m
    c. 125 kg
2. 221 cm

Homework

1. a. 24 L
    b. 324 L
    c. 224 L
    d. 575 cm
    e. 334 cm
    f. 365 cm
    g. 681 g
    h. 261 g
    i. 306 km
    j. 192 km

2. 174 g; tape diagram drawn and labeled to model problem
3. a. 158 min
    b. 19 min

Lesson 19

Problem Set

1. a. 280 cm
   b. 80 cm
   c. 365 g
   d. 254 g
   e. 648 mL
   f. 248 mL
   g. 4 km 233 m
   h. 2 L 51 mL

2. 149 km
3. 8 kg
4. 235 L

Exit Ticket

1. a. 159 m
   b. 108 kg
2. 78 kg

Homework

1. a. 190 g
   b. 166 g
   c. 287 cm
   d. 321 cm
   e. 842 g
   f. 542 g
   g. 2 L 20 mL
   h. 4 L 452 mL

2. 75 kg; tape diagram drawn and labeled to model problem
3. 188 km
4. 415 L

Lesson 20

Sprint

Side A

1. 200
2. 300
3. 400
4. 800
5. 1,800
6. 2,800
7. 3,800
8. 7,800
9. 300
10. 400
11. 500
12. 900
13. 1,900
14. 2,900
15. 3,900
16. 7,900
17. 500
18. 2,500
19. 400
20. 3,400
21. 700
22. 4,700
23. 400
24. 1,400
25. 500
26. 5,500
27. 900
28. 6,900
29. 600
30. 700
31. 700
32. 800
33. 900
34. 1,000
35. 1,000
36. 1,000
37. 10,000
38. 7,000
39. 4,100
40. 8,400
41. 3,600
42. 9,800
43. 2,900
44. 10,000

Side B

1. 100
2. 200
3. 300
4. 700
5. 1,700
6. 2,700
7. 3,700
8. 8,700
9. 200
10. 300
11. 400
12. 800
13. 1,800
14. 2,800
15. 3,800
16. 8,800
17. 400
18. 2,400
19. 500
20. 3,500
21. 900
22. 4,900
23. 300
24. 1,300
25. 400
26. 5,400
27. 800
28. 6,800
29. 600
30. 700
31. 700
32. 800
33. 900
34. 1,000
35. 1,000
36. 1,000
37. 10,000
38. 4,000
39. 2,100
40. 7,400
41. 4,600
42. 8,800
43. 3,900
44. 10,000

Problem Set

1. a. A: 295; 400, 200, 200
      298; 500, 200, (300)
      299; 400, 100, (300)
      302; 500, 100, 400
      B: 486; 700, 300, 400
      495; 800, 300, (500)
      498; 700, 200, (500)
      508; 800, 200, 600

   b. Explanations will vary; in the differences that gave the most precise estimates, both numbers either rounded down or both numbers rounded up.

2. a. Estimates will vary.
   b. 188 L; tape diagram drawn and labeled to model problem

3. a. Estimates and explanations will vary.
   b. 128 g; tape diagram drawn and labeled to model problem

Exit Ticket

a. Estimates will vary.
b. Estimates will vary.
c. 53 g
d. Estimates and explanations will vary.

Homework

1. a. 30 km
   b. 28 km
   c. Yes; it is a reasonable answer because our estimate is very close to our actual answer. A close estimate can help us see if our actual sum is reasonable.

2. a. Estimates will vary.
   b. 209 centimeters; explanations will vary.

3. a. Estimates will vary.
   b. 648 g

4. a. Estimates will vary.
   b. Estimates will vary.
   c. 254 liters of water; estimates and explanations will vary.

Lesson 21

Problem Set

1. a. 91 g, 58 g, 90 g, 60 g, 150 g;
       91 g, 58 g, 149 g
    b. 91 g, 58 g, 90 g, 60 g, 30 g;
       91 g, 58 g, 33 g
    c. Because both estimates are close to the actual answers

2. Yarn A: 64; 60
    Yarn B: 88; 90
    Yarn C: 38; 40
    a. Estimate: 100 cm; actual: 102 cm
    b. Estimate: 10 cm; actual: 14 cm;
       tape diagram drawn and labeled

3. Capacity of the 3 containers plotted and labeled on number lines
    Container D: 212 mL ≈ 210 mL
    Container E: 238 mL ≈ 240 mL
    Container F: 195 mL ≈ 200 mL
    a. Estimate: 650 mL; actual: 645 mL
    b. Estimate: 30 mL; actual: 26 mL;
       tape diagram drawn and labeled

4. a. 21 min
    b. Estimates will vary; actual: 94 min
    c. Because the estimate is close to the actual answer

Exit Ticket

a. Estimations will vary; 714 mL
b. Estimations will vary; 123 mL

Homework

1. a. Estimations will vary; 612 mL
    b. Estimations will vary; 306 mL
    c. Answers and explanations will vary.

2. a. Estimations will vary; 886 L
    b. Estimations will vary; 148 L

3. a. 26 min
    b. Estimations will vary; 11 min

4. a. Estimations will vary; 769 cm
    b. Estimations will vary; 312 cm;
       tape diagram drawn and labeled